Environment for Science

A History of Policy for Science in Environment Canada

Philip Enros

ISBN 978-0-9920704-0-3 (paperback)
ISBN 978-0-9920704-1-0 (ebook)

Library and Archives Canada Cataloguing in Publication

Enros, Philip Charles, 1950-, author
        Environment for science : a history of policy for science in
Environment Canada / Philip Enros.

Includes bibliographical references and index.
Issued in print and electronic formats.
ISBN 978-0-9920704-0-3 (pbk.).--ISBN 978-0-9920704-1-0 (pdf)

        1. Science and state--Canada--History. 2. Environmental
sciences--Canada--History. 3. Research--Canada--History.   4. Canada.
Environment Canada--Decision making--History.   5. Canada.
Environment Canada--Planning--History. 6. Canada. Environment
Canada--Management--History. I. Title.

Q127.C2E57 2013   509.71   C2013-905499-5
                                                    C2013-905450-2

Published by Philip Enros in 2013

Printed by CreateSpace, an Amazon.com company

CONTENTS

PART THREE: CAPACITY

PART FOUR: COLLABORATION

Illustrations

Acknowledgments

I began work on this history in July 2009 and finished three years later. During the first year, I was very fortunate to be able to work on it full time. The Public Servant-in-Residence Program of the federal government's Canada School of Public Service made that opportunity possible. Thanks to it, I spent the year at Carleton University in the School of Public Policy and Administration, which very generously welcomed me, giving me an office and library access. In addition, I am indebted to Environment Canada for allowing me to step away from my managerial responsibilities for the year while continuing to pay my salary and providing a space where I could review boxes of files. I hope that this book will realize the Department's investment in my project. For the two years after retiring from Environment Canada in 2010, I continued working on the book on a part-time basis. While at Carleton I had researched and written about the Department's policy for science activities after 1990, during retirement I dealt with the Department's first 20 years.

The book would not have been possible without the resources provided by several libraries and archives. In particular, I want to acknowledge the help that staff of the Environment Canada library, the Department's records management services, and Library and Archives Canada gave me in my research. I also wish to acknowledge the encouragement of my colleagues in the Science Policy Division. They patiently listened over many lunches to presentations of parts of my work and provided feedback. Several of them also gave me comments on a draft of the book. They nurtured my belief that a history of policy for science might be interesting and useful to others. Prompted by issues arising from my research, I contacted several former employees of Environment Canada when I had specific questions or needed orientation about events with which I was dealing. I want to thank them for their time and openness. Errors of fact or interpretation are, of course, entirely my responsibility. Last, but not least, I am happily indebted to my wife, Pragna, for her constant support and interest in my work.

The illustrations in this book, with the exception of those in the appendices, appeared originally in government publications or reports. Their copyright rests with Her Majesty The Queen in Right of Canada, with the exception of Figure 3.1 which is the property of Science-Metrix. They are reproduced here with the approval of the responsible departments or agencies and with the permission of the Minister of

Public Works and Government Services Canada, granted in 2013. Marc D'Onofrio of that Department's Crown Copyright and Licensing Office was very helpful in obtaining the permissions. My thanks are also due to Science-Metrix for allowing me to reprint Figure 3.1. The image on the book's cover was created using the concept cloud tool provided on the website *Text is Beautiful.*

I joined Environment Canada, in the Office of the Science Advisor, in 1992. During my years in the Department I had the pleasure of working on policy for science with and for many talented individuals. Through researching this history I have gained a better appreciation of the efforts of the many policy analysts and managers from before my time in the Department. This book is dedicated to all of those who have worked on policy for science in Environment Canada. Although their names, with a few exceptions, are not to be found in it, they are the ones who made this history.

Introduction

Science is the foundation of Environment Canada. These words have often been used by the Department to acknowledge the importance of science to its work. Environment Canada has always been one of the largest science-based departments of the Government of Canada. The majority of its employees serve in scientific and technical positions. The work they do accounts for more than half of its expenditures. Their activities allow the Department to fulfill its mandate by providing various environmental services to the public, developing and enforcing environmental regulations, and establishing sound environmental policy. Some of the scientific activity is devoted to research, which is mainly conducted to help improve the Department's performance.

However, just as the foundation of a building is largely invisible, Environment Canada's science and its contributions are generally forgotten as attention is focused on the purpose of the Department's activities, the environment. This often proved to be the case in the overall management of the Department. Taking care of the environment was without question its raison d'être. And that goal was very demanding of senior managers' attention. The management of science was commonly regarded as a less immediate priority. As a result, it often slipped to the margins of the Department's agenda.

Yet, science was and remains a core component of Environment Canada. This reality gave rise to a certain tension in the management of the Department. For example, in the midst of a major review in 1980 of how it managed science, the Department was cautioned about its tendency to neglect its scientific activities.

> It has been said frequently by senior management that scientific activities in the Department serve as tools to meet departmental objectives, and not for the pursuit of scientific knowledge for its own sake. Their crucial role in departmental affairs, however, requires that scientific activities receive much greater acknowledgement and attention that [*sic*] they have in the last few years. Like any good tool, science has to be used properly and looked after well.[1]

While science might not usually have been a top-of-mind issue, the Department's senior officials could not ignore its management.

This book recounts the history of Environment Canada's efforts to look after its science; that is, the story of its activities in policy for science.

---

[1] *Review of Science Management in DOE*, August 28 1980. EC Roots, box 8

It is not a history of the Department's scientific and technological activities.[2] Rather, it examines their strategic management – decision-making and commitments *about* science and technology. It looks at department-wide policy work, hence it reflects the preoccupations of the Department's senior managers. While policies developed by subdivisions of the Department or by the federal government sometimes enter the picture, they are not its focus. Nor is the book a history of the Department.[3] Its main goal is to make visible a single dimension of that experience: the Department's policy work in support of its science.

The book deliberately uses the term *policy for science* rather than the more generic *science policy* to try to avoid any confusion that it deals with the history of Environment Canada's use of science in environmental policy making. Advising on the implications of science for policy is one of the defining features of government science, in contrast to science conducted in a university or industry setting. It is one of the main functions of the Department's science. The use of science for policy is the subject of a very extensive academic literature.[4] In addition, several case studies have been done of Environment Canada's experience in this area.[5] However, that important topic is outside the scope of this book. It is only touched on in the final chapter, which looks at the Department's policy work on linking science to policy.

This history of policy for science in Environment Canada covers the 40-year period since the creation of the Department in 1971. It is organized into five parts. The first deals with where departmental policy work for science was done. Policy work requires sustained resources. How those were organized is an important element of the context for that work. The remaining four parts of the book look at the issues that were worked on, grouped under the themes of general strategy, capacity, collaboration and communications.

---

[2] A good starting point for exploring the Department's research is Brian Wilks, *Browsing Science Research at the Federal Level in Canada: History, Research Activities and Publications* (University of Toronto Press, 2004).

[3] A history of Environment Canada has yet to be written. A good introduction is G. Bruce Doern & Thomas Conway, *The Greening of Canada: Federal Institutions and Decisions* (University of Toronto Press, 1994). Several books have looked at the evolving role of the federal government in environmental policy and politics. See, for example, Kathryn Harrison, *Passing the Buck: Federalism and Canadian Environmental Policy* (UBC Press, 1996).

[4] To name only two recent works, see Roger A. Pielke, Jr., *The Honest Broker: Making Sense of Science in Policy and Politics* (Cambridge University Press, 2007) and Ann Campbell Keller, *Science in Environmental Policy: The Politics of Objective Advice* (MIT Press, 2009).

[5] See, for example, Debora Van Nijnatten & William Leiss, "Environment's x-File: Pulp Mill Effluent Regulation in Canada" in William Leiss, *In the Chamber of Risks: Understanding Risk Controversies* (McGill-Queen's University Press, 2001), 125-163.

Throughout its history, Environment Canada has always had an organizational home responsible for conducting policy work on science. However, that base has undergone several transformations. Part One tells the story of its evolution, which has experienced four main stages. The changes over time show the Department's struggle with the tension between managing science and environmental issues when making decisions about resource allocation. They also reveal the part played by a variety of forces – Deputy Ministers' views, pressures from the federal government's policies for S&T, the distribution of S&T across the Department, budgetary constraints and the experiences of its S&T workforce – within that setting, in shaping how policy for science would be organized at different points in the Department's history.

The resources devoted to the organizational base for Environment Canada's efforts in policy for science also reflected the importance and extent of the issues it was expected to handle. Those concerns were numerous and varied. The policy work on them has been arranged in this book into several clusters. The first, the subject of two chapters in Part Two, explores the Department's efforts to develop general strategies for its scientific activities. Chapter 2 examines work on overarching science strategies. Chapter 3 looks at strategies for two specific areas of S&T activity: Canada's North and international engagement. The motivation for developing strategies was varied, including asserting the importance of their subject, guiding decision-making, and establishing plans of action. The chapters show that the impetus for this policy work came both from within the Department and from other parts of the federal government. Much policy work resulted, but not many strategies. The challenges were significant. The interest by the Department or by the federal government in them proved to be weak. In addition, Environment Canada took few steps to integrate science strategies into its planning and reporting systems.

Obtaining the resources to conduct science is one of the main concerns of policy for science. Part Three reviews the history of the Department's policy work on S&T capacity. Its chapters explore the three components of capacity. Chapter 4 looks at efforts to make a case for funding for the Department's science. Chapter 5 deals with an important feature of government science, its physical infrastructure. In periods of tight finances, capital expenditures were usually the first to be cut in spite of the longer-term consequences. Chapter 6 focuses on human resource issues. Policy work here covered a broad range of issues bearing on the management of expertise, including recruitment, rewards, development and equity. Policy for S&T capacity not only had to reconcile competing interests within the Department, it also had to convince central agencies and Cabinet of the value of government science. The latter was often attempted in conjunction with other science-based departments. But, for most of the Department's history, this was a major challenge. The federal government usually struggled with deficits and was strongly lobbied to build up S&T capacity in the university and industry sectors. Policy work was necessary, but not sufficient, to deliver the needed resources.

Collaboration was another significant preoccupation of Environment Canada's policy for science. Part Four surveys the issues involved in partnering with the three most important science-performing sectors in Canada. Chapter 7 is devoted to policy for working with other federal science-based departments. Environment Canada's efforts were motivated by the crosscutting nature of environmental issues and by tight financial resources, which put a premium on leveraging and finding synergies in attempting to respond to a growing number of environmental priorities. The Department's experience here tells a great deal about what could and more often could not be done in the absence of changes to the governance machinery of the federal government. Environment Canada's policy work also sought to find ways to further develop environmental science capacity in universities, and to better align and link it to the Department's mission. Chapter 8 reveals the challenges the Department faced and its often-innovative approaches to getting more out of its scientific relationships with universities. Chapter 9 examines the Department's ambivalent engagement with industrial innovation over the years. Environment Canada did not wholeheartedly embrace industry. Its policy in support of environmental industry varied greatly over the years, due to the reduction of available funding and a consequent focusing on the Department's core activities. In addition, a narrow interpretation of its regulatory role in the last decade made a defence of a technology promotion role difficult.

Departmental policy around communicating its S&T is the subject of the last part of the book. Its chapters deal with two very different purposes for communicating. Chapter 10 looks at policy efforts to explain the role and value of Environment Canada's S&T to opinion leaders, central agencies, the media and the general public. This work was usually stimulated by external pressures such as criticism of the management of science, requests for reporting, budgetary cuts and, sometimes, negative publicity. It was pursued in order to correct what was perceived, by many in the Department, to be a widespread lack of understanding of the functions of its S&T and of awareness of its contributions. Chapter 11 examines efforts to improve the use of science in the development of policy and in decision-making. This was important to the work of the Department and to maintaining public trust. Work at the science-policy interface was a pervasive, ongoing and intrinsic part of the Department's activities. Most of the time, it went on without much notice as part of regular business. The chapter focuses on the exceptional period from 1997 to 2002, brought about by a controversy centered in two other science-based departments. Environment Canada's policy work on the science-policy interface was able to draw on its rich and extensive experience with the interaction of science and environmental policy.

A short epilogue and six appendices conclude this volume. The aim of the former is to summarize what the history presented in the various parts of the book tells us about doing policy for science in Environment Canada. The appendices hold a number of resources for the reader. Three of them provide data on expenditures, personnel and research publications. The others present a list of abbreviations, a note on sources and organizational names, and a timeline to help in positioning events.

The book is a history of Environment Canada's efforts in policy for science. By making that experience available, it transfers some of the lessons of the past to current and future science policy analysts, advisors and managers. A better appreciation of that history will, of course, be of interest primarily to those in government. But students of public policy and administration should also find this account of the practice of policy for science in Environment Canada instructive. In addition, I hope that the book will elicit the interest of historians and science studies scholars in exploring the field of policy for science. The book lays out the broad contours of Environment Canada's effort. The sources used by it illustrate the range and richness of the documents that scholars can use in making science in government more visible. There is little scholarly work in this area. Much more remains to be done, for example by looking at topics covered here in greater depth or at activities within the many branches of the Department. Similar studies for other science-based departments and agencies can also be carried out, providing useful comparisons to Environment Canada's experience. Additional scholarly context will, in turn, aid in obtaining a better understanding of Environment Canada's experience in policy for science.

Although Environment Canada has long seen science as its foundation and maintained a substantial scientific capacity, the issue of whether science should be done inside government has often recurred in public policy debates in Canada. This book, I believe, shows that the issue, as posed, is too narrow. It needs to be reframed. By telling the story of one government department's efforts in policy for science, the book deepens our understanding of the issues facing the management of directed science, the type of science needed by government. It shows that the roles and characteristics of the enterprise of science in the service of government have to be taken into consideration. The real public policy issue is how best to provide for the kind of scientific activity the government of Canada requires to deliver on its missions.

PART ONE

ORGANIZATION

1

Organizing Policy for Science

Environment Canada's efforts in policy for science date from its origins. Moreover, that activity was usually judged important enough to be given a prominent place in the Department's organization. Over the forty years since Environment Canada (EC) was created, several different approaches have been used to support, coordinate and carry out policy for science. This chapter looks at their history, and is structured around them. Its focus is on where department-wide policy for science was done in EC, its origins and evolution. The remainder of the book deals with the policy issues that were tackled by the people who worked in those organizations.

The Research Coordination Directorate was the first home for EC policy for science. It was short lived. The Directorate's founding idea – that coordination would get the most from EC's investment in science – proved to be unworkable in the early days of the large, diverse and decentralized Department. The Directorate was followed by the Office of the Science Advisor. This arrangement lasted for 20 years, although the Office underwent a number of changes during that time. A growing dissatisfaction with the overall management of science in EC led to a new approach. Policy for science, which for the first 20 years of the Department had been part of its policy structure, slowly shifted to being a responsibility of the Department's science operations, where it has since remained. The transition began in the mid-1980s. The outcome, a decade later, was a separate management system for science, managed as a distributed network led by the Assistant Deputy Minister (ADM) of the Environmental Conservation Service. After another decade, the system underwent a radical change; being streamlined and centralized under an ADM for S&T. Policy for science was a significant activity and organizational unit under both ADMs. All four approaches to organizing this activity, despite their differences, shared a common purpose. They provided a focal point for the strategic management of the Department's S&T capacity. They were an acknowledgement of the size of EC's science capacity and of its importance in achieving the Department's mandate.

That quite different mechanisms were tried reveals a deep management challenge with which EC has struggled over the years. The Department has had difficulty in finding a governance system that was able to devote sufficient attention to both EC's scientific foundations and its environmental mission, to both its means and its ends. Because S&T was usually regarded simply as a means to an end, its management tended to be overshadowed or ignored by the Department's administration. Pressing environmental issues largely monopolized senior managers' deliberations. Of course, other factors were also involved in the different responses to the governance challenge. The views of Deputy Ministers, pressure from the Federal Government's policies for S&T, the distribution of S&T across the Department, budgetary constraints, and the experiences of its S&T workforce all played a part in shaping how policy for science would be organized at different points in EC's history.

Research Coordination Directorate

When Environment Canada officially came into existence on June 11 1971, it included a unit, the Research Coordination Directorate, responsible for department-wide policy for science. Along with two other directorates, it reported to the ADM of Policy, Planning and Research. The intent was that it would work to ensure the coordination of the Department's research towards the Department's goals and would review the quality of that research.[1]

The Directorate was a new organization. Unlike most other units in the Department, it had not existed before the formation of EC. After the government had announced it would establish a new department in the speech from the throne of October 8 1970, several units from other departments were transferred to the Department of Fisheries and Forestry as part of the process of forming EC.[2] They were the Water Resources Sector from Energy, Mines and Resources, the Canadian Wildlife Service from Indian Affairs and Northern Development, the Meteorological Branch from Transport, the Air Pollution Control Division and the Public Health Engineering Division from Health and Welfare, and the Canada Land Inventory from Regional Economic Expansion.[3] This amalgam became known as Environment Canada when the Government Organization Act of 1970 received royal assent in June 1971.

The idea for the Directorate arose when the Department was being put together. In December 1970, the Minister of Fisheries and Forestry asked J. R. Weir to chair a small task force to consider and make

---

[1] Memo from DM, March 26 1971. LAC Acc. 1995-96/135, box 3

[2] The Speech from the Throne "proposed the establishment of a department to be concerned with the environment and the husbanding of those renewable resources that are part of and dependent upon it, with a mandate for the protection of the biosphere."

[3] All of the units, except for the Canada Land Inventory, were transferred by Order in Council P. C. 2047, November 26 1970. The Inventory was transferred later, in May 1971.

recommendations on an organizational component for the new department "that would achieve an integrated approach to environmental research".[4] Weir likely did not find the Minister's request perplexing. The need for better coordination of science as well as of natural resources was much discussed at the time.

The main stimulus for creating EC had been concern about pollution. But the design of the Department incorporated a broader perspective. It was a Department *of* the Environment. The management of environmental resources was an important consideration, as shown by the composition of the new Department. The nature of its activities in renewable resources meant that EC was very much based on science. When the Department was launched, it was responsible for the largest science expenditures in the federal government. It performed more than twice as much science as any other federal department or agency (Figure 1.1).

The state of Canadian science had been the subject of considerable attention at the time EC was formed. The same speech from the throne that had signalled the government's intention to establish the Department had also spoken of the "great wealth of untapped and uncoordinated scientific talent and experience not now adequately utilized in the quest for solutions to our modern problems." The Ministry of State for Science and Technology commenced operations just two months after EC. Its main purpose was to develop federal policy for science.[5] The formation of the Ministry had been preceded and probably shaped by two major reviews of federal science. The first volume of the Senate Special Committee on Science Policy (Lamontagne) appeared in 1970. And the Gendron task force, which had been asked in late 1970 to review federal science organizations, reported in April 1971.[6] In addition, the Science Council was publishing reports at the time on the need for greater coordination of research and of scientific activities in the area of natural resources. It recommended, in an October 1970 report, a Department of Renewable Resources to better integrate the work done by the federal government as well as an Environmental Council of

[4] LAC RG93 Acc. 84-85/81, box 3. Weir's credentials for this task were very strong. He had been an agricultural research scientist and a director of the federal Science Secretariat. In December 1970 he was chair of the Fisheries Research Board. In addition to research coordination, the task force was asked to make recommendations about an environmental advisory council.

[5] *Annual Report 1971-72*. Ministry of State for Science and Technology

[6] Andrew H. Wilson, "The Gendron Report: another view of Canadian science policy" *Science and Public Policy* vol. 16, no. 5 (October 1989): 269-281

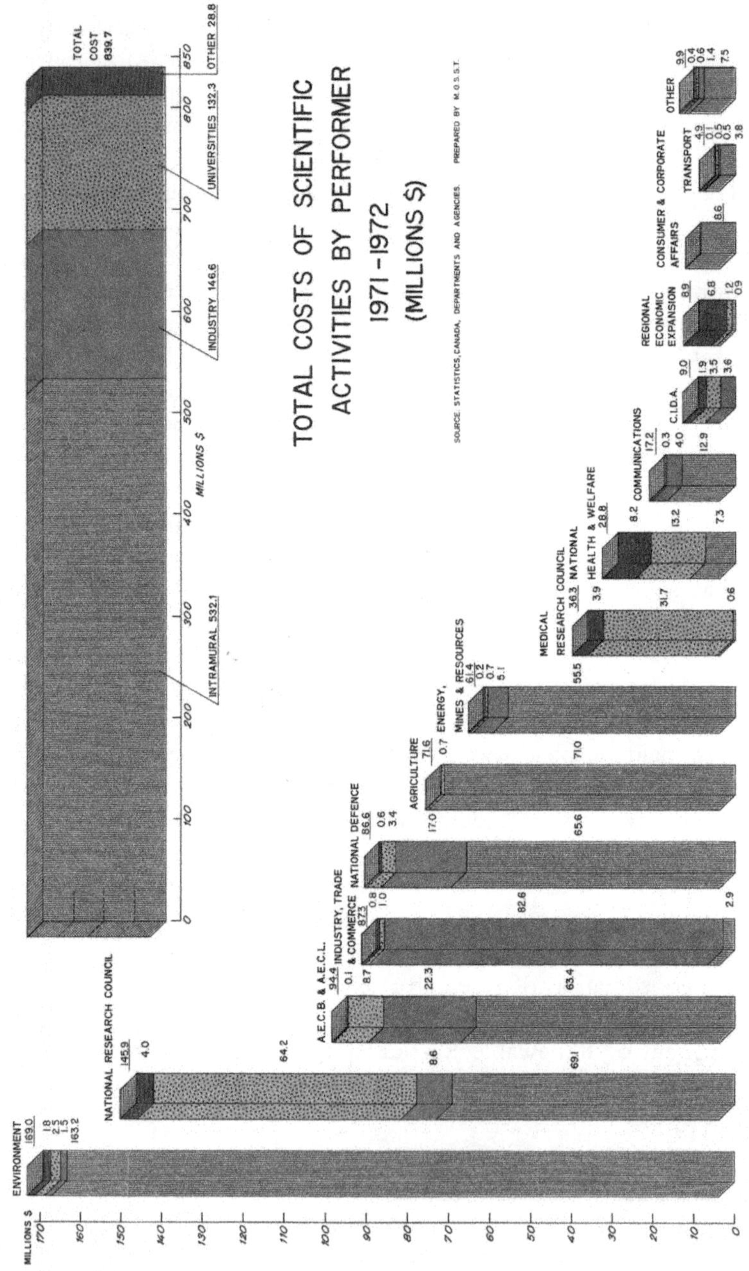

Figure 1.1 EC's first year science expenditures[7]

---

[7] *Scientific Activities: Federal Government Costs 1958-59 to 1971-72.* Ministry of State for Science and Technology, November 1971

Canada to conduct early assessments of environmental problems.[8] Given this background of public policy discussion about the need for increased coordination in Canadian science and in federal governance of natural resources, it was hardly surprising that the Minister asked for a unit devoted to integrating the research performed in the new Department's many and diverse technical services.

The Research Coordination Directorate consisted of 9 staff in 1971, mostly drawn from the Water Resources Sector that had been transferred to Fisheries and Forestry. Plans for the Directorate had it growing to 49 in two years' time.[9] However, like the rest of the new Department, the group had difficulties in staffing positions in the first year due to difficulties in obtaining Treasury Board approval.[10] Even the Directorate's head position, the Director General (DG), was vacant in the first year. It would only be filled in May 1972 with the appointment of a well-connected scientist from outside the federal government. Kenneth Hare took up the position through a new federal program, Executive Interchange, for a two-year term. Hare was a professor of geography and physics at the University of Toronto, a fellow of the Royal Society of Canada, and had been President of the University of British Columbia in 1969. Negotiations with him had been underway in November 1971 and completed by February.[11] It appears that he did not start until May 15th in order to complete his university duties for the term. While DG, Hare kept his doctoral students, who were working on projects directed by the Atmospheric Environment Service, and his external positions including director of Resources for the Future and governor of the Arctic Institute of North America.[12]

Although plans for the organization of the Directorate were frequently amended, the group was usually divided into two branches: science policy and ecological systems. The former had initially been named Research Strategies and Priorities, but this was changed to Science Policy in January 1972. Peter Meyboom was its Director. It was the group centrally concerned with policy for science and had most of the resources

[8] *This Land is their Land … : A Report on Fisheries and Wildlife Research in Canada.* Science Council of Canada, Report no. 9, October 1970
[9] In 1973 it had reached 29 person years, although 41 had been classified. LAC Acc. 1993-94/004, box 2
[10] Letter from Minister to the President of Treasury Board, January 21 1972. LAC Acc. 1985-86/342, box 2
[11] Hare recollected that he started to receive enquiries from Ottawa concerning what the federal government should do about its environmental mandate after the publication of an opinion piece by him, "How Should We Treat Environment?" *Science* 167 (January 23 1970): 352-355. See his unpublished autobiography *I Always Take a Window Seat! A Chronicle of My Life*, Trinity College Archives, University of Toronto.
[12] Memo from Hare to Armstrong, January 8 1973. LAC Acc. 1993-94/003, box 1

available to the Directorate. The Ecological Systems Branch was understaffed for the first few years. The idea was that it would focus on the life sciences, deal with environmental issues that were not within the domain of a single Service, and take a systems perspective.[13]

Despite a shortage of staff and the lack of a DG, the first year was very busy for Research Coordination in policy for science. The Directorate led or contributed to policy work on six topics that were taken to the departmental management committee. They were the design of a science subventions policy for the Department, a review of EC's scientific establishments, discussions about the implications of the new federal make-or-buy policy, about EC's northern science, and about its environmental monitoring, and a discussion paper on an overarching science policy for the Department.[14] The wave of interest in these topics was likely stimulated by the disposition by senior managers, during the initial year, to become more familiar with the various components and capacity of the Department.

The Research Coordination Directorate had its name changed to Research Policy in January 1973. The source of this change was the first major reorganization of the Department since it had been established about a year and a half earlier. Initially, the Department had been divided into seven Services: Environmental Protection; Atmospheric Environment; Water Management; Lands, Forests and Wildlife; Fisheries; Finance and Administration; and, Policy, Planning and Research. The main result of the restructuring was to divide EC into two operational components – the Fisheries and Marine Service, and Environmental Services – each headed by a Senior Assistant Deputy Minister. The Policy, Planning and Research Service was combined with the Finance and Administration Service to become the Planning and Finance Service, which included the Research Policy Directorate.

Hare was happy with the name change, writing that coordination "is not what we either try to or can do".[15] Coordination was a major challenge within EC. Right from its birth there had been a tension between developing a coherent approach to the environment and meeting the various legislative demands served by the disparate technical groups that made up the Department. In April 1971, the Deputy Minister (DM) had gathered his ADMs together in a retreat at Montebello to discuss the activities and priorities of the soon-to-be-launched EC. The meeting came up with six broad priorities, the first of which was to "carry

---

[13] LAC Acc. 1993-94/004, boxes 1 & 2
[14] Most of these topics are discussed in later chapters. Minutes of EC's Senior Management Committee, LAC Acc. 1991-92/017
[15] Memo dated January 15, 1973. LAC Acc. 1993-94/003, box 1

on established resource programs and services."[16]   It was a clear signal that the interests of the Services would place limits on departmental programs and policies.

Easing the tension between realizing departmental and service goals was likely behind the Directorate's name change. Hare reported in January 1973 that Meyboom "has had many problems with some services who challenge his jurisdiction and doubt whether a central science policy is needed."[17]   Despite his declaration about the Directorate's lack of a role in coordination, Hare did try to find agreement on what types of scientific coordination would be acceptable. On February 6th, he held a meeting with the DGs of EC's scientific units to discuss the issue. The key items on the agenda were the "limits, nature and mechanics" of coordination of scientific programs, the idea of Departmental science programs in addition to Service programs, and the role of the Directorate.[18]   The DGs "generally agreed that coordination based on identification of common scientific problems was needed."   Tellingly, there was no conclusion about the desirability of Departmental science programs. With regards to the role of his unit, Hare noted that it had no intention to run scientific programs but was "to be concerned with broad, general issues, and to propose to the management committee ways in which the department's resources can be brought to bear on problems."[19] The reaction to departmental programs showed that only weak forms of collaboration would be possible. The DGs, for example, did decide to continue to meet regularly to discuss science issues of general concern (the group became known as the Science Committee). Coordination of research would remain on the list of the Directorate's objectives, but it would not be realized in any significant program.

The Directorate had been established due to the notion that coordination would help to boost the impact of the Department's scientific capabilities. Coordination proved to be very difficult due to the autonomy of the Services and their focus on their mandates. The change in the Directorate's name signalled a retreat from its founding rationale. It marked a shift in the organization's mission away from departmental coordination toward providing advice to senior management. The scope

---

[16] The other five were focused on the environment: clean up and control pollution, assess and control the environmental impact of major development, promote and support international environmental initiatives, initiate long-term environmental programs, and develop an environmental information and education program. *Environment Canada: Its Organization and Objectives*, October 1971

[17] Memo. LAC Acc. 1993-94/003, box 1

[18] Memo, January 17 1973. LAC Acc. 1993-94/003, box 26

[19] Notes, October 1973. LAC Acc. 1993-94/003, box 26

of the Directorate's work on departmental policy for science, its main activity, was similarly constrained and shaped by EC's decentralization.

Office of the Science Advisor

Creation and First Five Years

A few months after the Directorate's renaming, it underwent a more significant change. In early March 1973, the Planning and Finance Service was reorganized, resulting in the breakup of the Directorate. Hare's DG position became known as the Science Policy Advisor and the rest of his staff, still mostly consisting of the Science Policy Branch, became part of the Policy, Planning and Evaluation Directorate. Hare continued to report to the Service's ADM, but now also sat on the departmental management committee and became associate secretary for the Minister's Canadian Environmental Advisory Council. He had no staff other than a secretary and administrative assistant. His role was essentially the same as before – working with the Science Committee, providing broad science advice, and maintaining links with the external scientific community – but without day-to-day managerial responsibilities.[20] Science Policy was kept as a Branch, continuing to focus on policy for science issues such as a Departmental science policy, the science subventions program, and obtaining information from other performers of science.[21] Peter Meyboom left his position as director at the time of the Directorate's dissolution. Marcus Hotz became the new director of Science Policy.

The separation of Hare and his staff did not last for long. In May, prompted by the departure of the DG of Policy, Planning and Evaluation, Hare was arguing for a regrouping.[22] By the end of August his staff had been reassembled into a new directorate called the Office of the Science Advisor.[23] As before, Hare was responsible for the Science Policy Branch and a second branch, called Environmental Systems. But the latter now had additional staff in the form of the Advanced Concepts Centre. It had been part of the design for Research Coordination in 1971, but had not been set up due to the staffing difficulties imposed by Treasury Board. The Centre's purpose was to take a broad and long-term look at environmental issues. It would operate mostly through contract

---

[20] Memo, February 28 1973. LAC Acc. 1995-96/135, box 104
[21] Memo, April 10, 1973. LAC Acc. 1993-94/004, box 2
[22] Memo, May 2 1973. LAC Acc. 1993-94/003, box 1
[23] Memo, August 28 1973. LAC Acc. 1995-96/135, box 104

studies.[24] The Centre became one of the foremost futures studies groups in the federal government.[25]

Hare would lead the Office for only a short time. His two-year interchange was due to conclude in May of 1974, but the University of Toronto was trying to get him back earlier.[26] In June 1973 the DM agreed that Hare could return in time for the fall term. He was given a personal service contract as a science counsellor to the Department, which turned out to be little used and was not extended at the end of March 1975. Before Hare left EC at the end of September, he had recruited Fred Roots to replace him as Science Advisor.[27] Roots was the Coordinator of the Polar Continental Shelf Program at the Department of Energy, Mines and Resources, a prominent arctic scientist, and heavily engaged in environmental issues.[28]

When Roots took up the duties of Science Advisor in early December 1973, the Office had 29 person years and a total budget of $855,000. About two-thirds of the resources were in the Science Policy Branch, with the remainder in Environmental Systems and the Advanced Concepts Centre.[29] Paralleling this organization, the Office was described as having three areas of responsibility: coordinating and assisting the Department's scientific activities in the areas of policy and of long-term and multidisciplinary work, providing support and advice on the scientific aspects of environmental and resource issues, and exploring future options and their social and economic consequences.[30] Since the Office was situated in headquarters, Roots saw it as serving the needs of the Minister, DM and other senior managers. Roots and his directors viewed the Office as a science unit, with highly qualified staff and an appropriate work agenda. The professional staff were expected to be working scientists with strong interests in environmental matters.[31] As the

---

[24] Note from Robert Durie to Fred Roots, February 7 1974. LAC Acc. 1992-93/252, box 28

[25] For an introduction to the futures movement in Canada see Fred G. Thompson, *Looking Back on the Future* (Futurescan Intl, 1992).

[26] Hare later commented that he was "fed up with the Ottawa scene." See his unpublished autobiography *I Always Take a Window Seat! A Chronicle of My Life*, Trinity College Archives, University of Toronto.

[27] Memo, August 17 1973. LAC Acc. 1993-94/003, box 1

[28] Roots had worked with Hare and Shaw on environment and polar regions aspects of Expo 67, had been involved in preparations for and participated in the UN Stockholm conference in 1972, and had been secretary to the 1971 Montebello workshop. Personal communication from Roots

[29] Memo, September 18 1974. LAC Acc. 1993-94/003, box 20

[30] *Office of the Science Advisor. Responsibilities, Activities, New Directions. Notes for Ministerial Briefing*, July 12 1974. LAC Acc. 1993-94/003, box 1

[31] Personal communication from Roots

majority of the Office's resources were in Science Policy, Roots set about to build up the other branches.

Policy work on science continued as a substantial activity. For example, Roots gave the EC presentation to the August 1976 meeting of the Senate Special Committee on Science Policy.[32] However, the efforts of the Office in other areas greatly increased. In January 1974 it was assigned responsibility as the Departmental focal point on energy. One of its staff, Bryan Cook, was sent on assignment as an EC contribution to the new interdepartmental office of energy R&D located at Energy, Mines and Resources. Other staff also worked on various aspects of energy such as conservation, renewable sources and nuclear. By 1977, the Office's work in this area had expanded to be a major part of its activities.[33] The Advanced Concepts Centre's activities also grew under Robert Durie. It was designed to operate with only a few staff and to contract out studies. Its efforts clustered around the subjects of futures studies, the conserver society, solar energy and environmentally appropriate technologies. In addition to these two major areas, the Office also worked on producing an environmental quality index and, through Brian Emmett, pushed for the incorporation of an economics perspective in the Department's environmental thinking and policy work.

Fulfilling Roots' concept of the Office would not be easy. In the first place, challenges in working with the Services continued. Roots was chair of the Science Committee. The DM had told him that it "was free to become what its members felt to be most useful."[34] But it was an opportunity that went unseized. The DG members preferred to meet informally as needed to discuss their problems without any specific reporting lines. They did not see the Committee as having management or operational responsibilities.[35] When Roots and his staff tried once more to promote the concept of Departmental science programs, the DGs had "many doubts as to the value and effectiveness" of the approach. Given this lack of commitment it is not surprising that the Science Committee met infrequently and disappeared after June 1975.

An additional challenge was that the Office's intended focus on science advice was being displaced by longer-term environmental policy work. The activities of the Advanced Concepts Centre had given the Office a capacity in that policy domain which proved useful for the Department. The Centre, assisted by others in the Office, had a major role in 1974 in preparing *A Perspective on the Next Decade for Environment*

---

[32] LAC Acc. 1993-94/003, box 27

[33] See the entries for the Office in EC's annual reports for this period.

[34] Minutes. Science Committee. January 15 1974. LAC Acc. 1993-94/003, box 26

[35] Minutes, Science Committee, February 24 1974. LAC Acc. 1993-94/003, box 26

Canada, which was the basis for the Department's ten-year plan.[36] A discussion by Roots and his directors about the Office's role in October 1975 noted that they were "being called upon increasingly to provide advice, not on scientific matters per se, but on policy directions and policy statements concerning the use of resources and environment by society."[37]

The final and main obstacle Roots faced was obtaining the resources needed to carry out the Office's work. Starting in his first year as Science Advisor, there were budget cuts and problems in acquiring new staff due to Treasury Board controls on person-years. At first Roots was successful in obtaining some additional staff through internal reallocation within the Service, which brought the Office up to about 35 person-years and a total budget of just over $1 million. But cuts in 1975 reduced the staff to 25. Roots reorganized the Office. As a consequence the name Science Policy was lost. The three branches were now Scientific Services, Integrated Programs and the Advanced Concepts Centre.[38] But this structure was short-lived. Under increasing budget pressure, the number of corporate policy staff in EC was cut by almost half in 1976 as the Department began a major zero A-base review. The Office was reduced to 14 staff with a budget of about $590,000. This reduction was accompanied by a change in reporting relationships. The Office as well as the policy side of the Policy, Planning and Evaluation Directorate now reported to the new Senior Assistant Deputy Minister of Environmental Services, Paul Tellier, who saw his role as including corporate policy.[39]

The Office's branch structure was abandoned. The professional staff now all reported directly to Roots. They were called advisors, of which three were titled science advisors with the remainder responsible for energy, economics, remote sensing and statistics.[40] Compounding the cuts, Roots had trouble keeping his staff. Some left the Office, others went on assignments within the Department, and a few left for international postings facilitated by the Office's good reputation.[41] For example, Durie went on a two-year interchange with the New Zealand Commission for the Environment in 1976 and Hotz took a two-year leave to work for the Scientific Affairs Division of NATO in 1977.

---

[36] *A ten-year action plan for Environment Canada,* 1974. On the influence of the OSA's document see a note from Munro in August 1975. LAC Acc. 1985-86/342, box 3
[37] Directors meeting, October 1 1975. LAC Acc. 1993-94/003, box 2
[38] OSA Directors meeting, October 29 1975. LAC Acc. 1993-94/003, box 2
[39] Memo from DM, January 18 1977. LAC Acc. 1993-94/003, box 4
[40] Memo, April 15 1977. LAC Acc. 1993-94/003, box 4
[41] Personal communication from Roots

Partition of the Office and the Science Review

The next major organizational change affecting EC's policy for science capacity was the creation of a separate Department of Fisheries and Oceans. The break-up had been foreshadowed in a series of events going back to the formation of EC. Much of the debate in the House of Commons at that time had concentrated on the implications for fisheries. Over the decade of the 70s, the political interest in environment had waned in favour of that for fisheries. EC had been divided into two components in 1973, one of which was the Fisheries and Marine Service. A Minister of State for Fisheries had been appointed in 1974. And the Department had been renamed Fisheries and Environment Canada in 1976. While the departure of fisheries removed significant resources, it also provided an opportunity to refocus attention on the Department's environmental mission.

The intention to create a Department of Fisheries and Oceans was announced to the departmental management committee by the DM in September 1977.[42] However, the new Department would not officially come into existence until April 1979. During the intervening 18 months, considerable discussion took place about the mandate of the federal government in the environment and the role of EC. For example the Senior Assistant Deputy Minister for Environmental Services (Jacques Gérin) established a writing team and a review group to draft a white paper on the environment in 1977.[43] Roots and his Office were major contributors to the effort.[44] The result of these deliberations was that the primary objective of the redefined EC was to be the preservation and enhancement of environmental quality.[45] A consequence of this renewed mandate was the transfer of Parks Canada from Indian and Northern Affairs to EC in June 1979.

The period leading up to the establishment of Fisheries and Oceans also saw much thinking in the spring and summer of 1978 about the role of the central policy unit in EC. As a significant element of corporate policy, the Office was very much involved. Roots soon became uncomfortable with the vision that was developing. He was concerned that his group would not have any real substance and therefore be unable to attract staff with the scientific and technical background needed to deal with environmental issues.

---

[42] Minutes, September 22 1977. LAC Acc. 1991-92/017, box 20
[43] LAC Acc. 1995-96/360, box 4
[44] EC Roots, box 10
[45] Memo, August 1978. LAC Acc. 1993-94/003, box 24

> In my more pessimistic moments I see no room in the
> 'central group' as it is evolving for the kind of approach,
> enquiry, and output that I have been trying to build in the
> Office of the Science Advisor over the past five years. For our
> size, we have been remarkably productive. . . . Hardly any of
> this could be done by a purely service unit, but hardly any of it
> fits into a purely policy-oriented central headquarters unit. Do
> I, as one of the central policy planners, willingly sweep all this
> aside, and with it the real job of a Science Advisor, without
> anyone really appraising the importance or effectiveness of
> what we have been trying to do?[46]

The design for the corporate policy function did not go Roots' way.
The Senior Assistant Deputy Minister favoured a plan for a group that
would assist the Departmental management committee.[47]

The debate about the role of a departmental policy group took place
in the context of pressures to reduce this function. Treasury Board
favoured smaller policy groups, and ongoing budget cuts as well as the
departure of Fisheries and Oceans were additional reasons for cutting
back. These pressures would lead to a reduction of EC's central policy
staff, which had already seen major attrition in recent years, from 73 to
about 40 person years. The impact on the Office was significant. As had
happened for a brief period in 1973, its staff was transferred to the policy
group of the new Corporate Planning Group. All that remained in the
Office was the Science Advisor, an assistant and a secretary.

The Science Advisor continued to be a member of the departmental
management committee and to be executive secretary of the Minister's
Canadian Environmental Advisory Council. Roots now reported to the
DM. At the same time he was supposed to work closely with the
Corporate Planning Group, headed by the Senior Assistant Deputy
Minister. Roots' mandate remained essentially the same. Besides reporting
to senior management on the implications of federal policies for the
department's science, he also was to provide scientific advice on energy
issues, on polar regions and on other issues that were not the
responsibility of existing programs.[48]

The impact of the Departmental reorganization of 1979 on policy for
science was a major reduction in the resources devoted to that area of
policy as well as a fragmentation in accountability for it. Roots had many

---

[46] Memo from Roots to Gérin, August 9 1978. LAC Acc. 1993-94/003, box 5
[47] Jacques Gérin, *A Central Function for the New Department of the Environment*, August 28
1978. LAC Acc. 1993-94/003, boxes 10 & 24
[48] Selected Activities of the Science Advisor, November 22 1978. LAC Acc. 1993-94/003,
box 24

other responsibilities and little assistance. Although the Office's staff had been transferred to the Corporate Planning Group, few remained there. The latter needed to bring in Rick Lawford from the Atmospheric Environment Service in October 1979 on a two-year assignment to work on policy for science, where he was a one-person operation. EC's interaction with the Ministry of State for Science and Technology was also divided. The ADM of the Atmospheric Environment Service was the usual EC representative on interdepartmental science ADM committees, and the Corporate Planning Group distributed material from the Ministry and provided departmental information to it.

The hollowing out of the Office of the Science Advisor occurred, ironically, at a time of increasing criticism of EC's management of science. The discontent was fuelled by several factors. First and foremost was pressure on the Department's budget. EC had undergone a major two-year Zero A-base review of all of its programs starting in 1977. Further reductions of over $40 million were announced in September 1978 leading to the termination of or cutbacks in about 20 programs. The second factor was a conviction that the federal government did not value its own scientific effort. The government's Make-or-Buy policy, an attempt to stimulate industrial research through outsourcing federal research, was often cited as evidence of this.[49] The final cause was a belief in a growing gulf between EC management and the Department's science, leading to a demoralized scientific staff.

> Those at the top of this department spend all their time standing in the turrets with machine guns defending attacks or pillages on departmental activities. There is a great void in the middle – void in the sense of an intellectual vacuum. At the bottom, there are many people engaged in carrying out programs but with a great sense of isolation and frustration, little sense of how they fit in the overall scheme of things.[50]

The unease about EC's management of science found expression in three reviews that were undertaken in 1979. The first was conducted by the Minister's Canadian Environmental Advisory Council. It was prompted by concern about the impact of the 1978 budget cuts on the quality of EC's science. The Council had raised the issue at a joint meeting with the departmental management committee on August 2nd.[51] The DM "welcomed the idea of a responsible investigation into the

---

[49] For more on the policy see chapter 4.
[50] Quoted by Roots in his "Introductory Notes to Workshops on Science Policy and Science Management" (1980). EC Roots, box 8
[51] LAC Acc. 1992-93/011, box 18

problems of maintaining effective scientific capability and productivity under present financial realities" and pledged departmental cooperation.[52] Council member Mervyn Franklin, President of the University of Windsor and a biologist, led the review. The Council submitted several questions to the Department concerning how its goals and priorities guided its scientific work, whether it had a policy to generate and share new knowledge, and how it set priorities for and evaluated its research.[53] Based on EC's responses as well as briefings and interviews with senior officials, Franklin produced a draft report in January 1980. After discussions by the Council, a revised version was accepted in June.[54] By this time, EC had already begun its own review, which the Council had been hoping to stimulate through its effort. So the Council was content at that time to summarize the report in a letter from its chair, Donald Chant, to the Senior Assistant Deputy Minister, Jacques Gérin.[55] Eleven "problems of science" were identified, the most serious being that "science in DOE lacks organization and focus, it is perceived as having low priority, and its fruits are not always seen as being influential in the setting of policy." Two key actions were recommended: appointing an ADM for science who would be an advocate for science but have no line responsibilities, and establishing a scientific advisory committee composed of senior scientists in EC. The Council then decided to step back and monitor what action the Department would take as a result of its own review.

The second assessment was a lengthy – approximately 70 pages – essay on the low morale of scientists in the Department's Environmental Management Service. Titled *A Scientific Scream: An Examination of the Environment for Science in EMS*, it was proffered as a "frank appraisal" of the "frustration, outrage, anguish, and cynicism of the scientific staff of EMS." This cri-de-coeur was written by L. C. Newman, a Departmental forestry research scientist on assignment as a science policy analyst in the Service's Policy and Program Development Directorate. His work, which appeared in October 1979, was based on group interviews with researchers and their managers held in the spring at several research centers. Although the essay was about one part of the Department, and even though Newman soon left for a job in the United States, it was widely circulated in EC and in other departments. It was used by the

---

[52] As reported by Roots on September 14 1979. LAC Acc. 1992-93/011, box 56

[53] LAC Acc. 1993-94/003, box 10

[54] *Research Policies and Priorities in the Department of the Environment. Final Report,* June 17 1980. LAC Acc. 1992-93/011, box 45

[55] Letter dated August 29 1980. EC Roots, box 10

Department in its efforts to strengthen communications between scientific staff and senior management.

A similar but much less personal account of the deteriorating work environment for EC S&T staff appeared the following year. Professor George Farris of York University was contracted by the Department to survey S&T employees for their reactions to the effects of financial restraint on various policies and practices.[56] Ninety-nine persons participated in the study, 40% of whom were research scientists. Overall, his study found a low level of satisfaction, confirming Newman's views.

The last of the examinations of science management in EC in 1979 was produced by the Office of the Auditor General. Its annual report, released in December, contained a section critiquing the Department's management of R&D.[57] The Auditor General found the "systems and procedures for planning, organizing and controlling" R&D weak. The report called for improvements in these areas as well as a consistent, department-wide standard for R&D management.

EC's senior management had not been unaware of the unease in the Department's scientific staff. The April 1979 meeting of the departmental management committee had agreed to look more closely at the idea of an annual review of the Department's science by the committee.[58] But the three reviews appear to have made the issue a priority. New resources were devoted to it. John Tener, a former ADM of the Environmental Management Service, was brought back from retirement in October 1979 as a special advisor to inquire into the management of scientific programs.[59] That same month, Rick Lawford was brought into the Corporate Planning Group on assignment for the purpose of management development. He spent all of his time on policy for science, being responsible for a project to identify policy and management issues related to EC's science. Based on the three reviews, work being done by MOSST and discussions with EC staff, that project identified 24 issues by December. The Corporate Planning Group thought that the three most important were procedures for setting R&D priorities, a Departmental science policy, and identification of core research activities.[60] Lawford also produced a lengthy paper, *Science and Technology Activities in DOE*, which presented data on S&T expenditures and human resources. Noting that the "future of R&D in the Department is unclear" and that current

[56] George F. Farris, *Some Effects of Administrative Policies on DOE Scientists*, 1981. EC Roots, boxes 7 & 8

[57] The section is in chapter 14 (which deals with EC) of the *Report of the Auditor General of Canada to the House of Commons: Fiscal year ended March 31, 1979*.

[58] LAC Acc. 1991-92/017, box 29

[59] Ibid., box 32

[60] Memo, December 14 1979. EC Roots, box 10

trends were disturbing, it also proposed several recommendations for moving forward, some of which would be acted upon.[61]

Fred Roots was also asked to prepare a discussion paper reviewing the "trends, problems and issues" facing science in EC, for presentation to the departmental management committee. His *Science in the Department of the Environment, 1980* was tabled at the January 24 1980 meeting of the committee along with extracts from Lawford's paper.[62] Roots noted the difficulties the Department had in dealing with science.

> There is a widespread feeling that DOE is not, as a Department, well equipped to answer questions or deal with issues in fundamental areas of this important component of the Department's activities and character.[63]

He recommended preparation of a report on EC's science.

Soon afterwards, the DM announced a comprehensive review of the issues affecting the management and use of science. Its purpose was to improve the Department's effectiveness in using its scientific resources, in providing financial and personnel support for science, and in exchanging and using science.[64] Regional workshops would be held. Tener would lead the exercise, assisted by Lawford. Roots was asked to work closely with them and to provide liaison with the Canadian Environmental Advisory Council whose own review was well underway at that time. Tener would also be responsible for preparing a report with recommendations for action for the September meeting of the departmental management committee.

In total, seven two-day workshops were held across the country involving 147 EC scientists and managers.[65] All but one were held in May, the last was in July for middle management in headquarters. The workshops revealed a general consensus across the Department about the issues facing its science. These were wide-ranging and included

> . . . the definition of DOE's goals, science policy, scientific leadership, the communication of policy matters, program evaluation, over-administration and under-management, the accountability of line managers, the authorities of line managers, management skills, DOE's scientific credibility, program planning, inadequate financial support, the

---

[61] EC Roots, box 8
[62] LAC Acc. 1991-92/017, box 33
[63] EC Roots, box 8
[64] Memo from DM to ADMs and RDGs, February 1 1980. EC Roots, box 10
[65] *Summary of Science Review Workshops Conducted in May 1980*, and *Report of the Headquarters Science Review Workshop* (July 1980)

implementation of budget cuts, program justification, personnel policies, classification systems, appraisals, the morale of scientists, vertical and horizontal communications, public information programs, Central Agency policies and demands, contracting out, administrative procedures, DOE's support to Universities, and DOE's organization.[66]

Based largely on the input from the workshops, but also on some other work done within the Department (e.g., by the Personnel Directorate), a Science Review Action Plan was prepared. It consisted of 26 recommendations. Each was discussed and action was approved at a special meeting of the departmental management committee on September 16 1980.[67] The actions covered five areas: the departmental mandate and science policy, program management, communications, personnel management, and relations with central agencies.[68]

In November, the DM sent a memo to all scientific and technical staff reporting on the output of the Science Review.[69] He began by commenting on the role of science in the Department.

> As I noted at the outset of that special Management Committee meeting, it is my firm conviction that the fulfillment of the Department's mission depends fundamentally on the quality of the scientific information we obtain and transmit. Our ability to learn and understand and to transmit that knowledge is essential to the discharge of our responsibilities. There is no question that in the last few years of restraint and transformation, the quality of our scientific effort has suffered. In addition, the urgent nature of many of the diverse initiatives brought before Management Committee has meant that we, as a group, have not had the opportunity to devote our full attention to the impact of those factors on science programs in the Department. This review and the recent meeting enabled us to focus directly on science as an activity and the problems facing scientists, rather than submerging science in the other issues of the Department.

The actions to follow up on the Review responded to issues of concern to the Department's scientists. They represented an attempt by EC to pay more attention to its scientific effort and its context. Interestingly, none of the actions were focused on the Department's

---

[66] Ibid.
[67] LAC Acc. 1991-92/017, box 35
[68] The actions and their follow-up are covered in other chapters.
[69] Memo, November 21 1980. EC Roots, box 10

organization for dealing with those issues. It appears that senior managers felt that the existing system – involving the Science Advisor, the Corporate Planning Group, and the departmental management committee – was sufficient. This belief would soon be challenged.

In Search of a Science Management System

Progress on implementing the Science Review was closely monitored for a year. Despite the fact that most of its recommendations were acted upon, a sense of unfinished business remained. The Canadian Environmental Advisory Council believed that more emphasis should have been put on the quality of EC's science. It shared its opinion with the DM in September 1980 that the Review focused more on the well being of the Department's scientists rather than on the well being of its science.[70] The DM did not disagree, but believed that "first we need to show the scientists we are taking their difficulties seriously."[71] After about a year, the Council decided to convey its views to the Minister by letter. It congratulated the Department for undertaking the review, but counselled that much remained to be done to improve the climate for research in EC.[72] The Council restated the two recommendations it had sent to the Department a year earlier – creation of an ADM for research without line responsibilities and of a science advisory council of leading departmental scientists – and added that the Department should better communicate the centrality of science to its mandate as well as put in place a regular system of peer review for science projects. The Minister (John Roberts), who also happened to be responsible for MOSST at that time, replied "I share your concerns regarding science, not only for the Department, but for science in government and industry as well."[73] However, his response simply described steps the Department was taking to follow up on the Review, a tacit support of its approach to science management.

Within the Department the lack of progress on some of the fundamental elements of the Review – e.g., a statement on Departmental science policy – contributed to the view that science continued to be neglected.[74] Roots described the mood as "a disheartened conviction among many that in a Department which depended for its information

---

[70] Chant to Seaborn, September 19 1980. EC Roots, box10
[71] Seaborn to Chant, October 24 1980. EC Roots, box 10
[72] Chant to John Roberts, June 1 1981. LAC Acc. 1992-93/011, box 45
[73] Roberts to Chant, August 6 1981. LAC Acc. 1992-93/011, box 45
[74] Report by F. Roots, *Maintenance and Utilization of Science in the Department of the Environment*, 1983

and effectiveness on vigorous and sustained scientific activity, scientific quality was not appreciated and the nurturing of science, as such, was not felt to be important."[75] A number of negative indicators were cited as marking a decline: the ongoing restraint in science expenditures both in the Department and in the government, a "perceived decrease in the quality of our science and scientists," a shift to more short-term work, an inability to keep up with advances in environmental science, a "lack of coherence and co-ordination of scientific activities amongst the Services," and the need to develop multidisciplinary approaches and deal with increasingly complex environmental issues.[76]

Contributing to the challenges in dealing with science management issues was the inadequate level of resources devoted to policy for science. The Corporate Planning Group relied on the efforts of Rick Lawford. He was not replaced when he left in the fall of 1981 upon completion of a two-year assignment. The Senior Assistant Deputy Minister, Jacques Gérin, decided not to staff any position in the Group related to policy for science. He wanted Roots to assume full responsibility for such work and suggested that he hire an assistant.[77] Gérin became DM at the beginning of 1983. He again urged Roots to devote more of his efforts to improving the state of science in the Department.

> I seek more impact, more influence on the mainstream of orientations, not on the margin (the things that others don't do) while I recognize the need for "freedom" and the fact that the very value of the job comes from its unstructured nature and the freedom to roam about. I also recognize the benefits that accrue to the department from your presence in a number of activities which would not otherwise be covered (though I continue to have doubts about that side of your occupations).[78]

Roots was reluctant to spend the majority of his time on policy for science. He saw his role as being an outward-looking science advisor.[79] He put greater priority on being involved in scientific activities, especially in the areas of energy and the North, and in the work of international scientific bodies.[80] Nonetheless, Roots did hire an Associate Science Advisor, Alan Cairnie, late in 1983. The Office also included a secretary

---

[75] Ibid.
[76] Memo, November 16 1983. EC Roots, box 7
[77] Memo, March 29 1982. EC Roots, box 7
[78] Memo from DM to Roots, May 27 1983. EC Roots, box 1
[79] Memo, May 16 1983. EC Roots, box 1
[80] See the entries for the Office of the Science Advisor in EC's *Annual Report* for 1980-81, 1981-82 and 1982-83.

and had a budget of about $200,000.[81] For the next three years, it would be the focal point for departmental policy for science. The DM continued to press the Science Advisor "to place greater emphasis on influencing the management of the department to come to grips with some difficult issues related to management of our science programs and our scientists."[82]

In September 1983, prompted by ongoing questions about the management of EC's science, Roots tabled another discussion paper at a meeting of the departmental management committee. Titled *Maintenance and Utilization of Science in the Department of the Environment*, it highlighted

> . . . some characteristics and problems of maintaining scientific capability and relevance in the Department, and suggests that more specific attention be given, than has been given in the past, to scientific activities as such. Such increased attention may lead, in the Departmental interest, to changes in Departmental policy, structure, or resource allocation.

Among those changes, Roots called for "more careful consideration in a context less defensive of the status quo, and in the light of changing external factors," of organizational alternatives for the management of EC's science, such as those previously recommended by the Canadian Environmental Advisory Council.[83] At the very least, Roots believed that there was a need for a headquarters unit to take responsibility for reporting on and coordinating science.

The departmental management committee agreed in November 1983 on a number of actions in response to the issues Roots had raised. Among them was the intention to organize a meeting between the committee and a group of EC senior scientists.[84] This did not happen. A year later, following a discussion of a draft EC statement on science policy, the committee decided to form an advisory group composed of senior scientists.[85] Its purpose was to advise the DM on science priorities and new areas of science requiring attention, as well as on how to improve the productivity of EC science and ensure that science was fed into management decisions.[86] The group had 17 members: all the senior research scientists (those at the top level of the RES classification), two senior scientists each from Parks and from those involved in

---

[81] EC Roots, box 8
[82] Memo from DM to Roots, March 1985. EC Roots, box 1
[83] *Maintenance and Utilization of Science in the Department of the Environment*, 1983
[84] LAC Acc. 1991-92/017, box 40
[85] Ibid., box 42
[86] EC Roots, box 4

environmental protection technology, and the Science Advisor and his Associate.

The Senior Scientists Committee had its first meeting only on May 2 1985.[87] The delay in launching it was probably due to various upheavals in the Department at that time. The Forestry Service had been transferred from EC to the Department of Agriculture in September 1984. In addition, the arrival of a new government under Brian Mulroney saw very controversial program reductions in EC. The minister, Suzanne Blais-Grenier, announced in November 1984 cuts totalling about $46 million to Parks, the atmospheric environment service, and particularly to wildlife research.[88]

The DM attended the full meeting of the Committee, which was held at the National Water Research Institute in Burlington. He commented that he felt "some chagrin that despite his eight years in the department, this was the first time he had met all the senior scientists, and he now found that they had not all met each other."[89] The meeting discussed current policy for science issues, including the Auditor General's report on research in the Inland Waters Directorate, the Wright report on federal policies and programs for technology development, and Roots' presentation to the Nielsen study team reviewing EC.[90]

By the time of the Committee's second meeting, in September 1985, the Department had a new DM, Geneviève Sainte-Marie. She attended part of the meeting. The DM expressed "a strong desire to make the Department work in a more unified way with stronger crosslinkages." She invited the Committee to give her its view on Departmental priorities, strengths and weaknesses in science, and the maintenance and enhancement of science capability.[91] It began work on a submission on science management.[92]

The DM launched a management review of the Department in the fall of 1985, largely in response to the Nielsen report of that summer. The review did not include science management. Despite her invitation to the Senior Scientists Committee, Sainte-Marie set up a separate small task force in November to look at the issue. The group consisted of the ADM

---

[87] Minutes of the meeting. EC Roots, box 4
[88] LAC Acc. 1992-93/011, box 55
[89] Minutes of the meeting. EC Roots, box 4
[90] See chapter 11 of the 1984 *Report of the Auditor General of Canada*. It was responded to by a consultants' study *Review of the Current Systems in NHRI & NWRI and Alternative Systems for Water Management Research* (Gore & Storrie, October 1985). The Nielsen study team was led by George Layt and was released as *Programs of the Minister of the Environment. A Study Team Report to the Task Force on Program Review* (1985).
[91] Minutes of the meeting, September 25-26 1985. EC Roots, box 4
[92] Minutes of the special meeting, November 6 1985. EC Roots, box 4

of the Corporate Planning Group (Robert Slater), his policy DG (Diane MacKay), the Science Advisor and his Associate. Their report was ready by early January 1986. It underlined the unhealthy situation of EC's S&T.

> S&T programmes are not as effective as they ought to be; S&T staff are unhappy; our S&T (with some exceptions) is not first-class; policy staff are frustrated with attempting to maintain a continuing, stable S&T programme with declining resources and progressively shorter-term need for justification; for most observers, environmental science does not, in itself, fit obviously into federal goals for science to increase industrial economic performance.[93]

The report also noted a number of external critiques, including a recommendation from the 1985 Nielsen study team calling for a stronger corporate-level capacity to direct the Department's science in coherent interdisciplinary programs focused on priority environmental issues. In addition, growing requests from the Ministry of State for Science and Technology for senior-level representatives and for input into government-wide initiatives such as annual federal S&T reports were highlighting gaps in existing departmental management structures.

For the authors of the report, the main reason for the current state of S&T was the narrow focus on ensuring that the S&T of each major part of the Department simply serve its own operational requirements. They called for a broader, coordinated cross-Service approach and put forward two options to deliver it. They considered the idea of an ADM for S&T, but rejected this as requiring a major restructuring of the Department. Instead, they recommended a corporate S&T group under a DG of S&T reporting to the ADM of the Corporate Planning Group, which would be concerned with strategic planning, policy and budget coordination for S&T. The main duties would include establishing an S&T strategic management plan and playing a leadership role in S&T policy development, interacting with central agencies, recognizing new technological opportunities and coordinating technology transfer programs, and coordinating cross-Service R&D programs.

The report of the task force did not lead to any changes in EC's science management. The DM was "open to having eventually a DG for science" in the Corporate Planning Group.[94] However, it did have an impact on the Senior Scientists Committee. The uncertainty created by

[93] *Management of Science and Technology in Environment Canada.* Report to the Deputy Minister from the Science Management Task Force, January 6 1986. EC OSA, box 1
[94] Minutes of the meeting of the Senior Scientists Committee, May 5-6 1986. EC Roots, box 4

the management review and by the task force meant that the Committee did not meet again until May 1986. That would turn out to be its last meeting. One planned for December 1986 was postponed. Some of its members felt that it was not "as effective or useful as we hoped."[95] The DM indicated her preference that the Committee not meet.[96] The experiment with an advisory committee of departmental scientists had ended.[97] The ADM of the Corporate Planning Group was calling for a "fresh start" in managing science.

The new approach began with a shift in responsibility for policy for science back to the Corporate Planning Group. Its ADM was given responsibility for dealing with MOSST. The latter was entering a period of renewed activity accompanied by increased demands on science-based departments. The ADM was supported in this task by the Energy Division, which began to provide policy services for S&T from the beginning of 1986. The Division was quite small – the core staff were John Hollins and Wayne Richardson – and could only spend a limited amount of time on policy for science. At the same time, the Office of the Science Advisor was again reduced to Roots and a secretary when Cairnie left the Department in June of 1986. Roots concentrated on his mandate as a science advisor, dealing with issues that were scientific and environment-related but which did not fit into the Department's existing operations. Reflecting the growth in its new function, the unit in the Corporate Planning Group changed its name in 1989 to the Science and Energy Division.

A major innovation in EC's science management occurred in the fall of 1986 with the creation of the Working Group on Science Management by the Management Advisory Board, a sub-committee of the departmental management committee.[98] The Group was charged with advising senior management on science management issues of concern across the Department. Phil Merilees, the DG of atmospheric research in the Atmospheric Environment Service, chaired the Group, which was composed of a small number of representatives of research centers, the Science Advisor and the staff of the Energy Division who provided the group's secretariat.

The Working Group on Science Management was preoccupied with the same kinds of challenges as the task force had been earlier that year. The issue of corporate science management was discussed at both the

---

[95] Ian Stirling, December 19 1986. EC Roots, box 4
[96] Memo from Roots, July 19 1989. EC Roots, box 9
[97] However, a group of the Department's senior scientists, calling itself the Senior Scientists Committee, was active a decade and more later.
[98] EC OSA, box 1. The working group was established at a time of major reorganization of the Department.

first meeting in November and the second in December. The Working Group believed that the corporate function in support of departmental S&T needed strengthening and decided to prepare a paper on the subject. Underlining its importance, the Deputy Minister attended the second meeting and told the group that the management of science was at the top of her list of priorities.[99]

The paper was quite modest in its approach.[100] It rejected the previous suggestions for an ADM for S&T or a DG for S&T in Corporate Planning as not feasible at that time, instead setting out what was considered to be the bare minimum machinery necessary (Figure 1.2).

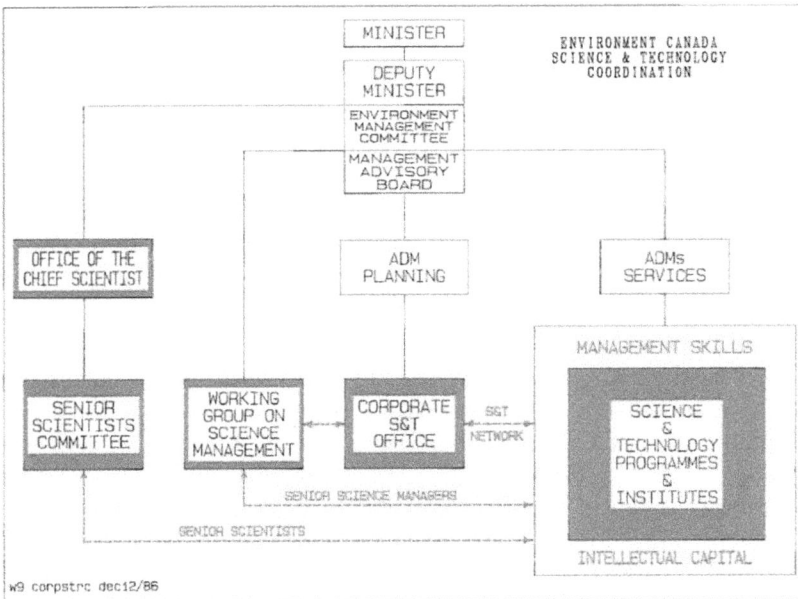

Figure 1.2 Proposed organizational chart for S&T coordination, 1986

The proposal built on and engaged the S&T and S&T operational management capacity within EC's Services by slightly adjusting existing committees and offices. In place of the Office of the Science Advisor, it

---

[99] Minutes of the Second Meeting, December 16 1986. EC OSA, box 1

[100] *Corporate Coordination of Science and Technology at Environment Canada.* Prepared for the Management Advisory Board by the Working Group on Science Management, January 21 1987. EC OSA, box 1

recommended an Office of the Chief Scientist. The new name signalled a desire to move the Office beyond merely an advisory role. It would be the focal point at the corporate level for EC's scientific community. The Chief Scientist would chair the Senior Scientists Committee and be the main source of scientific expertise and advice for the DM and the Department. The Working Group on Science Management would continue with its mandate of advising on Departmental science management issues. The Corporate S&T Office would be an expansion of the existing Energy Division, adding three staff to allow it to properly handle the volume of policy for science business.

The Working Group saw the corporate coordination of S&T as consisting of three elements. The first was directed within EC. Those tasks would include providing advice on the Department's S&T to senior managers as well as the Minister, analyzing and reviewing memoranda to cabinet, and annual departmental reviews and reports on S&T. The second dealt with those outside of the Department. Included here were relations with central agencies and other science-based departments, as well as representation and secretariat support at national and international S&T fora. The last element was communications, which consisted of the production, coordination and dissemination of relevant information on environmental S&T for diverse audiences. This task would be carried out with the assistance of a senior staff member from the Communications Directorate.

The Working Group's paper did not lead to any immediate change. For the next two years, it and the other related offices coped as best as they could. In 1988 the Group began reporting through the ADM of Planning and was re-christened the Environment Canada Science Committee. That same year, it and the Energy Division organized a series of six departmental workshops across the country to look at federal S&T policy and departmental strategy.[101] Although a report on these sessions noted "a general mood that a stronger scientific culture is required in the overall management of the department", there was no mention of a need for changes in the management of science at the corporate level.[102] The main challenge was a lack of resources. The Department's budget and its expenditures on S&T were not increasing. The Science Committee's members were all very busy managing their own scientific activities. And the Energy Division could usually only react to requests for assistance. The latter's director, John Hollins, noted that the Committee coped but

---

[101] *National Workshops on Science and Technology: Directions for Environment Canada*, September 1988. EC OSA, box 1
[102] *Report to CPG Management Board on Science and Technology*. Science and Energy Branch, January 23 1989. EC OSA, box 1

was "not strong enough to develop our own issues or prepare for issues that we anticipate." He went on to describe its situation as:

> The Environment Canada Science Committee is alive, but not particularly well. It provides useful liaison between senior managers of S&T, representation at MOSST/TB meetings, and authority to secure information. It does not have the capacity to do projects.[103]

Revival and Demise of the Office

In 1989, new windows of opportunity opened for the organization of policy for science in EC. There was renewed public interest in the environment and the government was planning on launching an environmental agenda. A new DM, Len Good, was appointed in May. He had a keen interest in ensuring that the Department's environmental mission was supported by a strong S&T foundation.[104] Fred Roots, who had been the Departmental Science Advisor since 1973, retired.[105] Ongoing discussions in July about a departmental strategic plan for S&T had led to the recommendation that there should be a new Science Advisor who would be a member of both the departmental management committee and the Science Committee.[106] In November Good appointed a new Science Advisor, Alex Chisholm. The latter was given ADM rank and the promise of a much-strengthened office. To help meet that goal, the Science and Energy Division was transferred into the Office of the Science Advisor. EC policy for science was once again centered in the Office and led by it.

Alex Chisholm was an appropriate choice for the position. He had risen through the ranks of the Atmospheric Environment Service as a researcher and gained a reputation as part of the Canadian negotiating team for the successful and widely lauded 1987 Montreal Protocol on substances that deplete the ozone layer. He had become DG of atmospheric research in 1987 and in 1989 was the chair of the Science Committee. Chisholm initially conceived of the Office as having a twofold role: to ensure the flow of timely, concise and credible S&T information to the department, minister and government; and, to ensure the health and national and international credibility of EC's

---

103 Ibid.
104 Personal communication from John Hollins, August 2009
105 He continued working for the Department as Science Advisor Emeritus, a position in which he was active for another 13 years before leaving in 2002.
106 *Vision, Values and Vehicles: A Strategic Approach for Environment Canada's S&T Programmes.* Workshop Report. Environment Canada Science Committee, July 1989. EC Roots, box 9

environmental S&T. To carry this out, he proposed an Office with a minimum of 12 staff.[107]

While Chisholm had been named Science Advisor in late 1989, it took several months and a number of exchanges with the DM before the Office defined its mandate and acquired further staff. In early 1990, Chisholm had increased his request for Office staff to 23 with an O&M budget of approximately $1.5 million. Much like the Office in the 1970s, Chisholm saw his as staffed with "experienced professionals, each responsible for broad scientific areas."[108] They would be organized into two sections, strategic and corporate (Figure 1.3). The former would formulate science policy and propose strategic plans for implementation of departmental S&T programs. The corporate wing would represent EC on interdepartmental committees on S&T, coordinate EC R&D initiatives and the production of S&T reports with the Services, and manage the Panel on Energy R&D program.[109] Overall, the role of the Office was to serve as a focal point for the department in matters relating to environmental S&T and to advise the DM and senior management on strategic scientific issues.[110] Expanding on the initial ideas about the Office, its responsibilities were now described as developing an S&T strategy, representing the department in broad S&T matters, facilitating the formulation of policy based on sound science, developing policies and procedures for the quality and relevance of S&T, and maintaining a network of external S&T contacts who could advise on a wide-range of environmental topics.[111]

Chisholm's plans for the Office were immediately overtaken by the need to develop the various science components of the Progressive Conservative government's new environmental agenda, the Green Plan. As measured by funding, S&T comprised about a third of the Green Plan. The Office became responsible for developing five major S&T components of the Plan, which consumed most of its effort for the next two years. Due to this responsibility, early ideas about the structure and functions of the Office were soon amended.

---

[107] *Proposal. Science Advisor and the Office of the Science Advisor*, September 1989. EC Roots, box 9

[108] Draft memo to the DM re structure and budget of the OSA for 1990-1991, January 17 1990. EC OSA, box 5

[109] *Office of the Science Advisor. General Task of Staff Members*, February 15 1990. EC OSA, box 5

[110] *Draft memo to the DM. Budget for Fiscal Years 1990-91, 1991-92. Office of the Science Advisor*, March 9 1990. EC OSA, box 5

[111] Ibid.

Office of the Science Advisor
Environment Canada

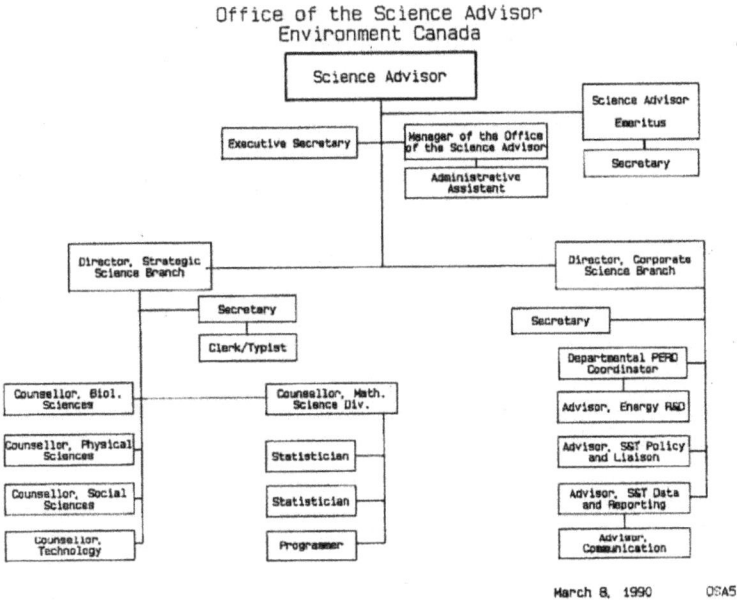

**Figure 1.3 Proposed organizational chart for the
Office of the Science Advisor, 1990**

For example, a new articulation of the mission of the Office was produced in early 1991 as part of a results definition exercise within the Department. While the mission – "provide strategic direction and leadership for the management of science and its input into environmental policy" – was essentially the same as that originally proposed, the needs served by the Office now included the Green Plan. The four needs were: effective management of EC's S&T; implementation and management of the Green Plan S&T Action Plan; ensuring that federal environmental policies and decisions are scientifically informed; and, supporting the overall development of environmental S&T.[112]

The four years in the life of the re-established Office of the Science Advisor were a very busy and challenging time. The resources given to the Office never reached the target Chisholm had set. In fiscal year 1991-92, the Office was allocated about half the staff (12) and O&M ($786,000) asked for the previous year. This grew to a staff of 20 and $3.4 million in O&M in 1993-94 due to the added responsibilities resulting from the Green Plan, but four of the staff and $2.8 million of the O&M

---

[112] *Results Definition Study. Office of the Science Advisor*, February 1991. EC OSA, box 5

were devoted to administering the Office's Green Plan S&T programs. Aside from its efforts to develop and implement those programs, the Office was also involved in many other activities, such as a departmental science forum, a biodiversity science assessment, the publication of an overview of the department's S&T and of a compendium of its R&D, and preparing a report on R&D decision-making in EC.[113] In addition there were continuing discussions about departmental management of science.

The revival of the Office of the Science Advisor had not put an end to concerns about how science was managed. If anything, it gave rise to and fostered further debate. There was an uneasy relationship between the Office and the Services. While Office staff were usually careful to involve the Services in relevant activities, this did not allay concerns about its role. When Chisholm presented his early 1990 conception of the Office at a meeting of the Environment Canada Science Committee, members expressed their concern that the Office "not compete with or overlap with activities of the Services."[114] Members were usually not shy about speaking out against perceived threats. When a proposal for a more integrated Departmental science policy was made at the Committee's meeting in June 1993, at which the DM was in attendance, the DG of atmospheric research asked "what had not been done before that a new integrated approach can do better?" and noted that atmospheric research was totally integrated into the Atmospheric Environment Service.[115]

The DM himself played a major role in stimulating dialogue about corporate science management. In March of 1991 he tasked the Office with preparing a profile of the Department's R&D and its management. The report back to the Deputy, in June, identified four management issues.[116] The first was a lack of cohesion, shown by a service rather than department focus and fragmentation in policy development, planning and management. The second was the lack of an explicit, departmental science policy and insufficient input of science into policy. Also needed was better management understanding of and support for science. The final issue involved the scientific community. There was a "syndrome of science as an underdog", an insecurity and defensive attitude. Scientists

---

[113] Some of these activities are discussed in other chapters.

[114] Minutes, Environment Canada Science Committee, April 10 1990. EC S&T Management Committee, box 1

[115] Minutes, Environment Canada Science Committee, June 21 1993. EC S&T Management Committee, box 1

[116] *R&D Profile and Issues Concerning the Delivery and Management of Science*, prepared by the Cybertec Consulting Group with the Office of the Science Advisor, June 10 1991. EC OSA, box 7

needed to better understand policy issues and be sympathetic to management needs.

In response, the DM asked the Science Committee and the departmental management committee for concrete proposals to improve the situation.[117]   This led to a number of activities, including a departmental Science Forum in 1992 and a report on R&D decision-making. It also gave life to another report on the structure of EC S&T management. This one proposed a new S&T Committee to "develop strategies for the optimal deployment of EC financial, human and knowledge resources, while providing an effective challenge function and a longer term perspective."[118] The purpose was to provide a "strong and coherent voice" for the Departmental S&T community to senior management on the organization, management and funding of S&T. The Committee would operate through six subcommittees: an executive, one on R&D (which would replace the existing Science Committee), and others on strategic outlook, quality, information transfer and monitoring. The overall theme of the report was the need to increase coordination in the management of the Department's S&T.

Not surprisingly, the proposal generated extensive discussion at the June meeting of the Science Committee at which the DM participated. Although he was of the view that the proposed committee should not be seen as a threat and that accountability for science across the Department was missing, the meeting ended with an agreement that more work was needed.[119]   Chisholm's advice was to abandon the idea of a new committee and to build instead on the existing Science Committee. Its membership and role would be expanded to reflect the content of the proposal. He believed it was now time to focus on collective accomplishments, not on "more conceptual discussions."[120]

These changes would not take place under the guidance of a Science Advisor. A new DM, Nick Mulder, was appointed on June 25 1993. He was not keen on advisory functions with limited management functions.[121] Other forces also conspired against the Office, including

---

[117] Minutes, Environment Canada Science Committee, July 3 1991. EC S&T Management Committee, box 1

[118] *Management of Science and Technology. Draft Proposal,* June 15 1993. EC OSA, box 3

[119] Minutes, Environment Canada Science Committee, June 21 1993. Interestingly, a consulting group, SECOR, had been asked by the Associate Deputy Minister to comment on the proposal. Their reply, which only came in July, viewed the proposed committee as tantamount to the creation of an ADM Science, with the complexity of not having line responsibilities over resources but being able to override priorities and decisions of line managers. Letter dated July 8 1993. EC S&T Management Committee, box 1

[120] Draft memo to the DM, June 25 1993. EC OSA, box 3

[121] Minutes, Environment Canada Science Committee, November 2 1993. EC S&T Management Committee, box 1

ongoing budget cutbacks, the Department's task force on administrative streamlining, the Clerk of the Privy Council's review of Deputy and Assistant Deputy Ministers, and the major changes in Environment Canada due to the decision by Prime Minister Kim Campbell to re-organize the federal government that summer. That decision extracted a major part of the Department, the Canadian Parks Service, causing a significant re-shaping of what remained. In the change, the Conservation and Protection Service was split, with Robert Slater named as Assistant Deputy Minister for Environmental Conservation. The Office of the Science Advisor was closed. In its place, a Science Policy Branch was created in the Ecosystem Conservation Directorate within the new Environmental Conservation Service.[122] Most of the staff of the Office were transferred to the Branch, with Chisholm becoming its director.

ADM Environmental Conservation

The demise of the Office of the Science Advisor marked a major shift in organizational responsibility for policy for science within EC. From the beginning of the Department, that function had been part of or associated with the departmental policy group. Now one of the science Services was in charge. The ADM for Environmental Conservation was given responsibility for the functional leadership of science policy across the Department. The shift also meant a loss of profile for the organization of policy for science and for the Department's S&T. The Science Policy Branch reported to a DG who was responsible for a number of programs, as was his superior, the ADM.

However, the closing of the Office did not have much of an impact on EC's system for dealing with policy for science. The ADM took over the key responsibilities of the Science Advisor, the Science Committee continued to meet, and the staff of the Science Policy Branch carried on with the work they had been doing in the Office. For example, the Science Committee met shortly after the announcement of the re-organization. It was now chaired by the ADM for Environmental Conservation, supported by the Science Policy Branch, and the DM attended part of the meeting. In a well-established pattern for DMs, Mulder acknowledged the importance of science to EC and declared that he was not a scientist. He did see four roles for the Committee: to determine S&T priorities, to identify where EC policies were going off-

---

[122] Message from the DM to all employees, October 1 1993. EC OSA, box 6

track and where they should be going, to determine how to manage science, and to be a forum for issues of scientific importance.[123]

Not only did the policy work go on, it began to greatly increase early in 1994. In February the new Liberal government announced a number of reviews, including a Program Review and a Federal S&T Review. At the same time the Office of the Auditor General was conducting an audit of the management of federal and departmental S&T, which was published later that year.[124] All of these involved EC's policy for science capacity. In addition, to get out ahead of these various reviews, the new DM, Mel Cappe, asked Slater to conduct a review of EC's S&T. The result was a report three months later, which identified seven S&T management issues raised by the ongoing Program and S&T Reviews.[125] One of the issues was direction setting. Here the report recommended a review of the operations of the science management and policy advisory mechanisms.[126]   After a discussion at the departmental management committee (then known as the Environment Management Board), the DM asked for an action plan for implementation of the recommendations. This was delivered in November and adopted by the end of that month by the management committee.

The *Action Plan for Managing Science and Technology at Environment Canada* set out 20 actions over an 18-month timeline. One of them dealt with rejuvenating the S&T management system. A new component was added to the existing system, an ADM-level S&T Executive Committee to bridge S&T managers and the Environment Management Board. The Science Committee, which had already been broadened beyond research managers by the addition of representatives from EC's five regions, was asked to establish a series of working groups on priority issues. It would soon be further expanded and renamed the S&T Management Committee. The Plan also recommended that mechanisms in the Services and regions be strengthened and better linked to the Committee. All of these changes were to be put in place by March of 1995.

The implementation of the *Action Plan* took place in very challenging times. In February 1995 the decisions taken under Program Review were announced. The number of employees in EC would be reduced by 25%

---

[123] Minutes, Environment Canada Science Committee, November 2 1993. EC S&T Management Committee, box 1

[124] *Report of the Auditor General of Canada to the House of Commons*. 1994. Volume 6, chapters 9, 10, 11 and 32

[125] *Science and Technology in Environment Canada. Report to the Deputy Minister*, September 22 1994. EC OSA, boxes 4 & 7

[126] The other issues were sustainable development, ecosystem approach and ecosystem science, partnerships, science for decision-making, commercialization, and people. Another issue, science for services, was soon added.

(1400 staff) over the next three years, and the Department's budget would decline by 32% (from approximately $726 million). These very large reductions took up a great deal of senior management attention and caused major challenges for the Department. The impact of Program Review would be felt by the Department for several years to come.[127]

Program Review delivered another blow to the organization of policy for science in EC. Most of the staff of the Science Policy Branch received separation notices. Alex Chisholm left on a special assignment before retiring. John Hollins took over as director. He also departed, after a year, on a pre-retirement assignment. Duncan Hardie then became director. The Branch was demoted to a Division for a short time. It seems likely that it would have been completely closed if not for the press of interdepartmental S&T policy issues. It became a Branch again in 1996, but with a bare minimum of staff. By 1997 it had five employees and had to make difficult choices about its priorities. This soon increased to eight and then to a dozen by 2002, at which level it would remain for the next ten years. Philip Enros became director in 2001.[128] Under him, the profile of the Branch staff shifted from one where they primarily had experience in departmental science programs to one where they were policy analysts who were also trained in science. All staff focused solely on policy for science, providing professional S&T policy services in support of EC's efforts to strategically manage its S&T. In December 2002, the Directorate in which the Branch was located was dissolved as part of a reorganization of the Environmental Conservation Service. As a result, the Branch reported directly to the ADM.

The S&T management system had stabilized by 1996 into a form that would remain essentially the same for the next eight years. As shown in Figure 1.4, the system consisted of three main components, supported by an S&T Secretariat linked to other S&T groups, departmental issue tables and the departmental management committee.[129] The S&T Executive Committee's task was to ensure a coherent approach to the management of EC's S&T. It was the main link to the departmental management committee, and directed the work of the S&T management system. Its membership consisted of the three ADMs with substantial scientific capacity – Conservation, Protection and Atmospheric – and one Regional DG. The Committee's chair was the ADM for Environmental Conservation. The S&T Management Committee's mandate was to support the Executive Committee by building consensus and developing advice on the strategic direction and management of S&T. It also served

---

[127] See chapter 4 for further information on Program Review.

[128] I was director until 2009, at which time Eric Gagné took over.

[129] Minutes, S&T Executive Committee, May 1996. EC S&T Executive Committee, box 1

as the vehicle for implementing changes to the management of EC's S&T. Its membership consisted of S&T DGs nominated by the three ADMs, the directors of all of EC's R&D centres and the director of the Review Branch. The chair was the DG responsible for the Science Policy Branch within the Environmental Conservation Service. The last component was the S&T Forum. Its mandate and membership were both very broad. It was meant to foster exchanges and share S&T, better link S&T and policy, identify new strategic issues, build consensus on scientific issues, and develop clear advice to support the S&T management system. Its members included all the members of the S&T

EC S&T Management System

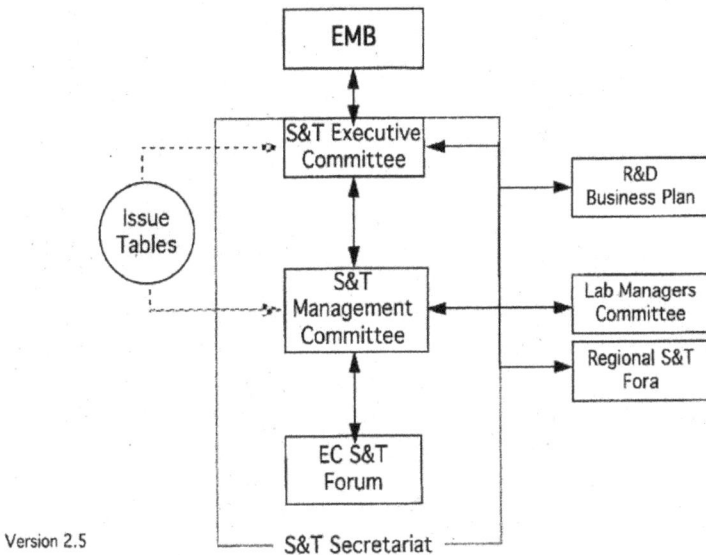

Figure 1.4 EC S&T Management System, May 1996

Management Committee as well as chairs of the lab managers' committee and regional S&T tables, staff engaged in the R&D business plan, leaders of issues tables, and representatives of the policy community. Because of this large mandate and membership, the Forum hardly ever met. Instead the S&T Management Committee would be enlarged over the years to include relevant elements of the broader EC community. Supporting all three components was an S&T Secretariat. Its core and lead was the Science Policy Branch, which worked with a network of individuals in the

Services and in the regions to support the meetings of the committees and to carry out the required policy research and services.

This S&T management system would remain largely the same for almost a decade. The following organizational chart (Figure 1.5) shows the system in 2003. In a department where S&T was distributed among several ADMs, the system enabled S&T managers to meet, discuss common issues and take action on them. During this period, the Executive Committee met over 40 times and the S&T Management Committee held almost 60 meetings. A considerable amount of work was done under their direction (some of which is covered by other chapters). The challenge in such a distributed system was obtaining the regular participation of managers, the timely input from the many groups involved, and maintaining the necessary links with the rest of the Department's management system.

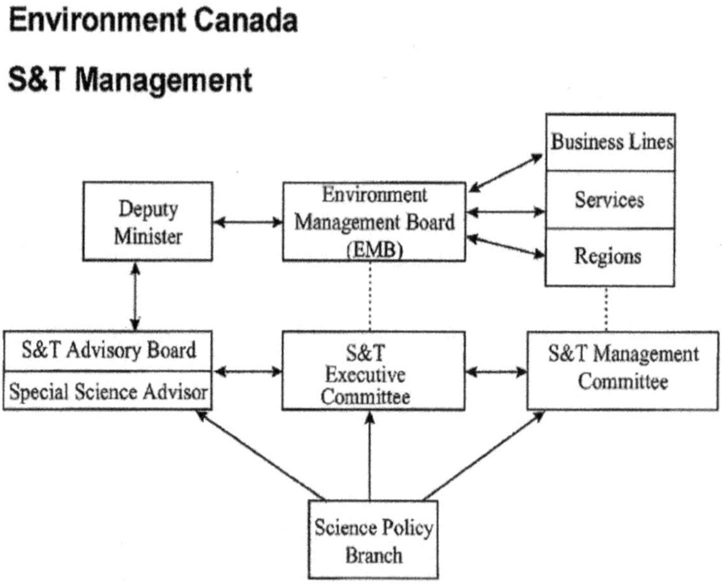

Environment Canada

S&T Management

Figure 1.5 EC S&T Management System, May 2003

The only major alteration to this S&T management system was the creation of an external R&D Advisory Board in late 1996.[130] The stimulus for this came from a commitment in the 1996 federal S&T strategy that all science-based departments and agencies would have

---

[130] For information on the Special Science Advisor, see chapter 8.

external advisory bodies to assess the relevance of their S&T.[131] In EC's case, the Board consisted of 11 members chosen for their experience and expertise. It met twice a year and reported to the DM. Its mandate was to "provide the Deputy Minister with broad, strategic advice on the relevance of the Department's entire R&D portfolio," to provide advice on plans and large integrated programs such as EC's business and strategic plans, and to respond to requests from the DM.[132] The Board's first meeting was on April 15, 1997 and was attended by the DM, as was almost every meeting after that.

For its first three years, the Board devoted itself to five areas which the Department considered to be high priority: science advice, science communications, R&D priority setting, environmental S&T capacity and integration of social sciences into the Department's planning and policy development.[133] Each topic was studied by a working group composed of some Board members as well as EC staff. In addition, two Board members served on the federal Council of Science and Technology Advisors, which was established in 1998, also in response to the 1996 federal S&T strategy. Council members were drawn from the various departmental S&T advisory bodies. Its purpose was to provide advice to the federal cabinet on the strategic management of S&T performed by the government. Over the next 8 years, the Council undertook a number of important studies of S&T in the government of Canada.[134]

Following an evaluation of its work, the R&D Advisory Board underwent several changes in 2000.[135] Its name had already been altered the previous year to the S&T Advisory Board, because many of its issues went beyond R&D. Now, the Board moved away from working groups producing reports, to a more direct advisory function to the DM. Membership and the frequency of meetings were increased. For the next five years, the Board met about three times a year, giving advice on such topics as biotechnology, environment and health, and climate change. The last meeting of the Board was in December 2004. Meetings were no longer called because the DM (Samy Watson) intended to replace the

---

[131] *Science and Technology for the New Century: A Federal Strategy.* Government of Canada, March 2006

[132] *Achievements to Date, Recommendations for Future Action.* Environment Canada. Science and Technology Advisory Board. Report No. 1, March 2000. Appendix B

[133] Ibid. The work of the Board on some of these topics is touched on in other chapters of this study. See also *Reinforcing External Advice to Departments*, report of the Council of Science and Technology Advisors, May 2001.

[134] In 2007 the Council and two other federal S&T advisory groups were replaced by the Science and Technology Innovation Council.

[135] *Review and Assessment of Environment Canada's Science and Technology Advisory Board.* Environment Canada, Review Branch, January 2000. EC S&T Advisory Board, box 6

Board with an environmental advisory body which would have included S&T. This plan did not come to fruition.

ADM Science and Technology

After more than a decade of stability, EC's system for dealing with policy for science underwent a radical change in 2005 with the creation of an ADM for S&T. The impetus for the shift did not come from any prolonged discussions within the Department. It was entirely due to the vision of the new DM, Samy Watson, who started his tenure in May 2004. Before coming to EC, Watson had been DM at Agriculture and Agri-Food Canada and had been responsible for a major re-organization there. It was clear from the beginning that he was determined that the Department would have an S&T Branch, headed by an ADM.[136]

The day following his arrival at EC, he told his management team of his intention to establish an external panel to give him advice on the management of the Department's S&T, something he had also done at Agriculture and Agri-Food Canada. Terms of reference were prepared that month and members selected by the DM. The Panel's task was to:

> … undertake a review and make recommendations for improving how Environment Canada manages its S&T resources and expertise to address departmental and government priorities. In particular, the panel should focus on the efficiency and the integration of the department's S&T expertise and the extent to which Environment Canada achieves this synergy across and beyond the department.[137]

The Panel met in July and completed its report by November.[138]

The report was structured around four themes: governance, quality, partnerships and emerging issues. Tellingly, the report did not make any recommendation about creation of an S&T Branch or ADM of S&T. It limited itself to making a case against having a chief scientist. It appears that the members knew of the DM's intentions or at least did not want to tie his hands in this matter.

The report did contain 21 recommendations, which were discussed by the Department's management in early 2005. Some initial steps were taken to respond to the recommendations, but these fell by the wayside

---

[136] Under Watson's changes, Services were renamed Branches. What were previously called Branches became Divisions.

[137] Terms of Reference, June 2 2004

[138] *Science and Technology Management Review Panel Report*, November 2004

as attention was increasingly devoted to other elements of the Department's restructuring.

There was considerable discussion in the summer of 2005 about what should be included in the S&T Branch. With the arrival of Brian Gray as ADM of S&T in August, he too became engaged in shaping the Branch.[139] When it officially came into existence in the fall, the Branch consisted of most of the Department's R&D capacity, its analytical laboratories, as well as its science and risk assessment functions. The Branch also included a new S&T Strategies Directorate to which the Science Policy group was transferred. The S&T Branch amounted to about one-fifth of the Department. For the first time, the Department now had a focal point for S&T whose primary task was S&T leadership.

Restructuring EC's major divisions was only one aspect of the DM's changes. More significant was a reversal of the previous dominance of the cross-cutting Departmental business lines by the Services. Under Watson, the business of the department, including budget allocations, was to be managed by cross-cutting Boards organized around intended results and collectively directed by the ADMs. Before, an ADM had been head of his or her Service and chair of the business line that corresponded most closely to that organization. Now, no single ADM chaired a Board. They were meant to take an active involvement in the business of the whole Department.

Since S&T was not seen as a Departmental result, it continued to be spread over the operational Boards. At that level, the management of S&T continued to be fragmented despite the unifying effect caused by the formation of an S&T Branch. Management structures continue to evolve. After the departure of Watson in May 2006, subsequent DMs tried to make the system work with only minor alterations. However, in February 2012, the DM announced a new governance structure. The Boards were abolished and the Branches regained authority over their operations.

Watson's changes did have a significant impact on EC's system for dealing with policy for S&T. With practically all of the Department's research and about 30% of its other scientific activities now reporting to one ADM, there no longer was any need for the cross-departmental S&T Management and Executive Committees. With the formation of the Branch on the horizon, these committees had ceased meeting by late 2004. In their place were the ADM of S&T and his management team. In addition, the science policy function was moved from the nature business

---

[139] Brian Gray was EC's first ADM of S&T. Trained as a wildlife ecologist, he came to the Department from Ducks Unlimited Canada on an executive interchange agreement. He left EC to be ADM of Earth Sciences in Natural Resources Canada in January 2011.

line (led by the ADM of Environmental Conservation) to the Strategic Integration Board, which was responsible for department-wide policy direction and advice. This re-established a tie with the departmental policy group and facilitated, in intention at least, greater interaction between corporate policy, focused on the environment, and policy for science.

The creation of the S&T Branch also made certain types of policy work much easier. A good example is the preparation of *Environment Canada's Science Plan*.[140] The Department had tried developing a longer-term S&T strategy before, for example with a research agenda exercise over the two-year period 1999-2001. But this could only achieve a synthesis of individual business line strategies.[141] With the Branch in place, a strategy was developed in one year that was a truly integrated document. The *Plan*'s three strategic directions transcended and were meant to apply to any part of the Department's S&T.

The two years under Samy Watson's direction saw the greatest change to the department's management since 1993. The bringing together of the Department's R&D in the S&T Branch caused a major change to the system of corporate management of S&T which had been in place since 1993 and which had reflected the distributed nature of S&T in EC.[142] This led to a much simpler, linear process for dealing with policy for science. However, because of the emphasis placed on Board management in EC, the formation of the Branch did not have a similar immediate impact on a more coordinated approach to the investment in and use of the Department's S&T by the results management Boards.

---

[140] Published in 2007. For more on this see chapter 2.
[141] See the draft synthesis, *Environment Canada Research Agenda*, October 2002. This was not implemented.
[142] By creating an ADM for S&T, Watson inadvertently fulfilled the 1980 recommendation of the Canadian Environmental Advisory Council.

PART TWO

GENERAL STRATEGY

2

Developing Policies and Strategies for Science

Just months after its establishment, EC was taking steps toward the development of an overall policy for its science. Though that effort was unsuccessful, the issue of the need for such a policy, or general strategy, for the Department's science would recur periodically, resulting in further work. This chapter examines the history of that policy effort.

The issue arose over the years due to changing departmental needs and circumstances. EC's initial effort was done to help guide decision-making on its science in the context of its then new environmental mandate. The second was pursued in order to affirm the role of science and raise its profile within the Department. Subsequent attempts focused on planning and priority setting, largely shaped by increasing government demands for accountability. The most recent work on a departmental science strategy, the *Science Plan*, stemmed from the consolidation of the Department's research capacity in the form of a Science and Technology Branch.

EC's policy work in this area was also prompted by federal government initiatives. For the first two decades, the Department responded to requests from the Ministry of State for Science and Technology for information on its scientific activities and for explanations about their function. This demand helped the Department develop a coherent picture of its S&T. The demise of the Ministry diminished the federal government's already weak interest in its own S&T. Since then, subsequent governments have issued several federal S&T strategies. EC participated in their development, when asked, and tried to influence them to support its S&T. However, the latter effort was not successful. The fixation of federal policy for science on the economy resulted in the strategies having little bearing on EC.

The importance of S&T to EC provided the basis for the various attempts at an overarching policy for its science. However, those efforts had little impact on the Department's regular processes for strategic planning and reporting. These concentrated on the environmental aims of the Department. S&T was regarded as merely the means to those aims. The lack of ongoing internal demand or use for science strategies meant that they were easily neglected. Central agency requirements for departmental science strategies and reporting would have changed this situation, but they never became entrenched. As a result, EC policy work would only occasionally be devoted to departmental science strategies, the Department would not have any regular means of articulating or profiling the strategic management of its S&T, and the development of overall policy for science would not be integrated with departmental planning.

Responding to EC's Needs

The First Step

In October 1971, four months after the establishment of EC, Peter Meyboom proposed drafting a discussion paper on the major science policy issues facing the Department.[1] At the same time, an article appeared in the Canadian magazine *Science Forum* reporting that although EC's Services were able to link their S&T to their planning and programs, this was not yet the case at the departmental level.[2] The ADM of Policy, Planning and Research (Al Davidson) was quoted as saying "we don't yet have an overall science policy." Meyboom's paper was intended to be the first step in the formulation of one.

Meyboom was the director of the Research Strategies and Priorities Branch (soon to be renamed the Science Policy Branch) in the Research Coordination Directorate. He had encountered a number of science policy issues during a study he had conducted on new research facilities for EC. He proposed to examine the issues and to suggest ways of dealing with them, in the hope that the result would provide guidance in departmental decision-making on science. His offer was accepted. Three months later, his paper *Science in a Changing Environment – proposals for a departmental science policy* was ready for the departmental management committee's review (Figure 2.1).

The paper's focus was on the new Department's environmental goals – the "new direction of the Department and its new scientific thrust."[3] It began by examining the broad governance challenges EC faced in responding to these new concerns through existing organizational units and in managing the institutional tendency to regionalization. It then looked at the role of scientists in government, distinguishing their work from that done by scientists in universities or in industries.

> It seems that for most of us the style of work should be neither abstract – academic, aimed at other scientists – nor

---

[1] Memo to A. T. Davidson, October 14 1971. LAC Acc. 1993-94/004, box 1

[2] Peter Calamai, "Building a new bureaucratic behemoth: Environment Canada," *Science Forum* issue 23, vol 4, no 5 (October 1971): 10-12

[3] *Science in a Changing Environment – proposals for a departmental science policy*, p. 21

service-industrial – aimed at industry – but pragmatic and down to earth, aimed at the public and its elected representatives. After all, work in government is a public service, and it is the public that has to be informed about the environment, the public and the press, because it is only through public concern that government measures can be taken, and only through credibility with the public that government measures will have effect.[4]

The paper made a policy proposal that government scientists working on environmental matters should consider the Canadian public as their primary audience.

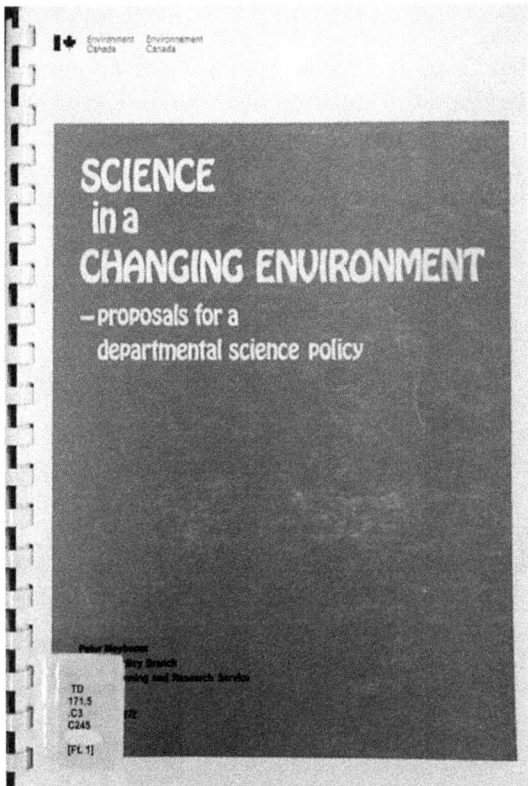

Figure 2.1 Cover of *Science in a Changing Environment*

---

[4] Ibid., p. 16

Science in a Changing Environment also laid out the kinds of concerns that flowed from the Department's environmental goals and their implications for EC's scientific activities. In addition, it examined how scientific programs were formulated in the government, proposed a methodology for setting priorities for research, suggested approaches to dealing with the need for interdisciplinary work in environmental science, projected funding requirements for the Department's science for the next five years, and offered up views on the design of regional research centres and on relations with universities, industry and citizen groups.

The paper was considered at the departmental management committee meeting on January 12, 1972. The committee felt that further discussion was needed. It asked Meyboom to review the document with a few senior managers from each Service and scheduled a special meeting for February 9th to continue its deliberations. There was considerable discussion at that meeting.[5] The management committee decided to distribute the paper and the initial set of comments, which Meyboom had obtained within EC. It also decided to send out an accompanying questionnaire to the approximately 2300 professional scientists in the Department. About a quarter of them replied. The results revealed the differences among the Services in their scientists' views on the issues covered by Meyboom. The responses to the questionnaire were also circulated by the management committee as *Science in a Changing Environment. Part 2: Interpretation of the Science Policy Questionnaire.*

Science in a Changing Environment would go no further. The same forces that had blocked the interest in coordinating EC's science were active here. The paper reflected the early excitement and ambitions in departmental headquarters about establishing a department of the environment. As such, it focused on the environmental goals of the Department and neglected, as its critics noted, the organization's work on renewable resources. The dominance of the Services in the early years and their close hold on the management of their science meant that there was little internal support for a departmental approach. Policy work on individual issues would go on, but there would be no further work on an overarching policy for science for another eight years.

Policy Statement on Science

By 1980, the demand for a policy statement on science had grown greatly in EC. The DM, in a memo to all scientific staff reporting on the results of the Departmental Science Review, noted that the formulation of such a statement was one of the Review's key recommendations and would

---

[5] LAC Acc. 1991-92/017, box 3

provide "more stability and consistency to the direction and management of science activities in the Department."[6] The Science Advisor was assigned the lead in preparing it.

The major factor in reviving interest in a department-wide policy for science was the criticism of the Department's handling of science, which had come to a head in 1979 and had led to the Science Review. In its listing of the policy for science issues facing EC in December 1979, the Corporate Planning Group ranked the need for a "framework for developing science policy" as one of the top three priorities.[7] This assessment was confirmed during the consultation process for the Review within the Department. It was not surprizing, therefore, that preparation of a policy statement on science was one of the follow-up actions from the Review approved by the departmental management committee in September 1980, appearing as the second item in a list of 26 actions.[8]

In addition to the Science Review, another government-wide development underpinned the call for a policy statement on science. Central agencies were increasingly pressing for more open and transparent management practices, a demand that included the management of research. For example, the Office of the Comptroller General's Improved Management Practices and Control study in 1979 resulted in a more systematic approach to program planning within EC and in other departments.[9] When the applicability of these practices to research efforts was questioned, an interdepartmental ADMs committee was formed to look at the issue.[10] The result was a draft document from the Comptroller General entitled *Operational Control of Research and Development in Departments and Agencies of the Federal Government of Canada* (1981). This was followed two years later by the *Framework and Guidelines for the Management of Research and Development in Departments and Agencies of the Federal Government of Canada*. It sought to "promote creativity while establishing standards of accountability for the resources invested in the R&D function." The activities of the Comptroller General along with the criticism of EC's management of research by the Auditor General in 1979 and the start of a study by MOSST on measuring the state of federal science that same year, contributed to the receptivity within EC for a policy statement on science.

---

[6] Memo, November 21 1980. EC Roots, box 10

[7] Memo, December 14 1979. EC Roots, box 10

[8] Minutes of the Special Management Committee Meeting, September 16 1980. LAC Acc. 1991-92/017, box 35

[9] EC *Annual Report 1980-1981*

[10] The ADM of the Atmospheric Environment Service was EC's representative on the committee. Letter from the Comptroller General to EC's DM, June 24 1982. EC Roots, box 9

The Science Advisor got off to a quick start on his assigned task. In the fall of 1980, he produced a document outlining the contents of a departmental science policy.[11] He also explained the need for such a policy. The lack of a shared view on EC's scientific work and its objectives among those who did the work, managed it and assessed it, meant that they lacked "a common basis on which to assess performance, need for resources, or relevance to Departmental priorities and responsibilities." He also noted that without a policy, decision-making about the allocation of resources for S&T was ad hoc and full cooperation with other performers of S&T difficult. Roots concluded:

> It is therefore important that the Department develop clear, explicit, widely discussed and openly available policy statements regarding the role of science in the Department of the Environment, and the policies of the Department with regard to the conduct of its scientific activities.

Several drafts of the policy were prepared and circulated over the following year.[12] However, the project soon fell by the wayside. It was difficult to develop a policy that addressed the issues at a department-wide level with the Services arguing for their unique circumstances. There were also many other issues requiring attention, and few resources to devote to them. For example, the ongoing work of the Office of the Comptroller General caused, in 1981, the drafting of a *Research Policy for the Department of the Environment* (which was rejected by research managers), the organization of two workshops for those managers to discuss work being done by the Office and by EC to improve R&D management, and the preparation of *Planning R&D Activities in Environment Canada* which built on the Office's work.[13] Roots also had several other priorities. The DM soon advised him that the statement on science policy could wait.[14]

However, the issue once again became a priority two years later. The renewed interest took place in the context of a strengthening concern that science in EC was not receiving the attention it warranted, despite the work done in the Science Review. The ongoing erosion of the Department's research capacity drove the dissatisfaction. The lack of

---

[11] *Science Policies in the Department of the Environment.* EC Roots, box 10
[12] F. Roots, *Maintenance and Utilization of Science in the Department of the Environment,* discussion paper, September 1983. EC Roots, box 8
[13] EC Roots, box 9
[14] Roots' introductory remarks, Science Policy Workshop, Strathmere, Ontario, January 31-February 2 1984. EC Roots, box 5

progress on a science policy statement was regarded as symptomatic of the neglect.

The event that would bring about a statement was a meeting in July 1983 between the DM (Jacques Gérin) and the Minister (John Roberts). It was prompted by an issue about conference travel by scientists. But it turned out that the Minister was more interested in being assured about the relevance of the Department's scientific effort.[15] As follow-up to the meeting, the DM requested a statement of EC's major research strategies for the Minister. In addition, the meeting reminded the DM of the commitment in the Science Review to develop an EC science policy. He told Roots that it was "now time for us to deliver on what we told scientists we would do three years ago."[16]

In response, Roots proposed having a discussion at a forthcoming meeting of the departmental management committee on the situation of science in the Department. He prepared a paper raising the issues surrounding the maintenance and utilization of science, which was sent to the committee in September.[17] The paper was discussed at the November meeting. One of the resulting action items was another request that the Science Advisor prepare a statement on science policy.[18] Based on his earlier work and on some new consultations, Roots was able to send a draft statement to the committee in July 1984.[19] Entitled *Policy Respecting Science and Technology*, it was short – only 4 pages – with another 10 pages of appendices (Figure 2.2). The core of the document was its first 1½ pages, which began with a concise statement of the policy:

> In fulfillment of its mandate, and in support of its objectives and specific policies, the Department of the Environment shall undertake activities and initiatives that will provide leadership in environmental science and technology in Canada, and enable Canada to play an appropriate international role in the environmental field.

The *Policy* then elaborated on this by stating that EC would undertake S&T in areas of its mandate for five reasons: to provide up-to-date information about the environment; to identify, assess and deal with environmental issues; to communicate S&T information about the

---

[15] Memo, DM to Roots, July 12 1983. EC Roots, box 7
[16] Ibid.
[17] *Maintenance and Utilization of Science in the Department of the Environment*, discussion paper, September 1983. EC Roots, box 8
[18] LAC Acc. 1991-92/017, box 40
[19] Among the new consultations was a science policy workshop held at Strathmere, Ontario on January 31-February 2 1984. EC Roots, box 5

environment; to encourage and facilitate environmental S&T capacity in Canadian industry, universities and other government agencies; and, to contribute to the growth and exchange of science. Roots' intended that the general policy statement would be fleshed out and applied as appropriate in each Service.[20]

The document's other significant components were contained in the appendices. One was a set of 10 principles for application of the policy. The other was a lengthy list of actions the Department and the Services could take to implement the policy.

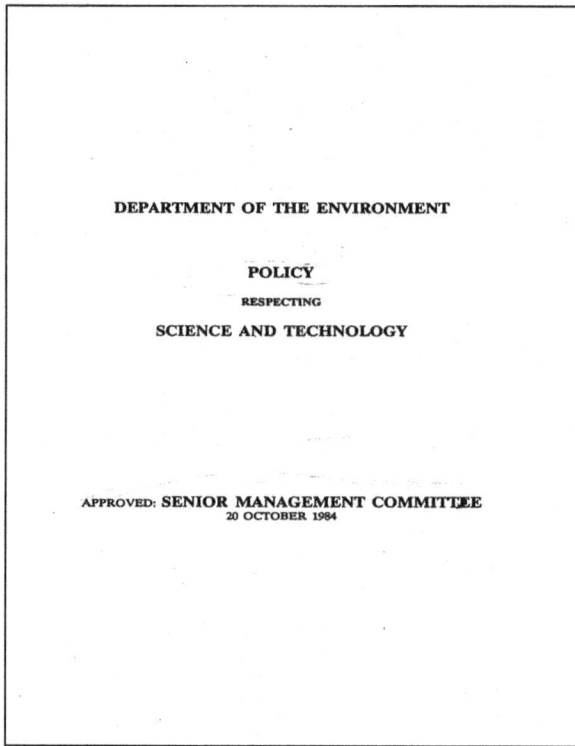

DEPARTMENT OF THE ENVIRONMENT

POLICY

RESPECTING

SCIENCE AND TECHNOLOGY

APPROVED: SENIOR MANAGEMENT COMMITTEE
20 OCTOBER 1984

Figure 2.2 Cover of *Policy Respecting S&T*

The *Policy* met with general approval at the departmental management committee meeting in October 1984. However, the document was stillborn, as the committee decided to delay its release. The timing was not good. A new government (Progressive Conservatives under Brian Mulroney) had been elected the month before, and a new Minister

---

[20] Memo, September 24 1984. EC Roots, box 8

recently appointed. The committee was concerned about releasing a policy affirming EC's scientific effort during a period of transition. Indeed, the following month saw the announcement of significant cuts to the Department's budget, including its scientific activities.

There were some attempts in the following 18 months to have the *Policy* distributed in the Department. The Office of the Science Advisor developed an implementation strategy for it in September 1985.[21] The short-lived Senior Scientists Committee recommended adoption of the strategy.[22] In March 1986, in an attempt to tack onto a recommendation from the Science Management Task Force, the Office recast the *Policy* into a strategic plan for science.[23] The latter proposed moving forward with a number of action items from the *Policy*, and was again supported by the Senior Scientists Committee. However, the DM (Geneviève Sainte-Marie) did not agree to distribute it.[24] Policy attention shifted to looking for new departmental management systems for science and to obtaining increased resources for EC's science.

Setting R&D Priorities

While the *Policy Respecting Science and Technology* soon faded from corporate memory, the issue of a science strategy did not. For example, the Department's Science Committee held an S&T strategy workshop in July 1989 that recommended an annual S&T business report as well as a set of criteria for making strategic choices in environmental S&T.[25] One of the *Policy*'s principles had stated that the requirements of science be considered as an integral part of the Department's planning and review process. Preparations for the Green Plan opened up new opportunities for realizing this goal. An emphasis on better integrating the Department provided an occasion for S&T to be a clear part of EC's planning and reporting.

---

[21] *Implementation Strategy for the DOE Science Policy*, discussion paper by A. B. Cairnie, September 11 1985. EC Roots, box 8

[22] Minutes, November 6 1985. EC Roots, box 4

[23] *Management of Science and Technology in Environment Canada.* Report to the Deputy Minister from the Science Management Task Force. Draft, January 6 1986. EC OSA, box 1. *DOE Strategic Plan for Science. Outline*, March 18 1986. EC Roots, box 4

[24] Minutes, Senior Scientists Committee meeting, May 5-6 1986. EC Roots, box 4

[25] *Vision, Values, and Vehicles: A strategic approach for Environment Canada's science and technology programmes.* Workshop report, 1989. EC Roots, box 9

Faced with the imminent announcement of the Green Plan and its ensuing operational responsibilities, EC's senior managers gathered at Montebello in July 1990 to discuss the state of the Department and the challenges before it. They characterized EC as a "department under stress".[26] Its external environment included new and complex environmental issues, intensified public concern, increased public expectations and steadily diminishing financial resources. Internally, the Department lacked a clear identity, cohesion and the confidence of many of its staff. EC was perceived as being in a downward spiral (Figure 2.3).

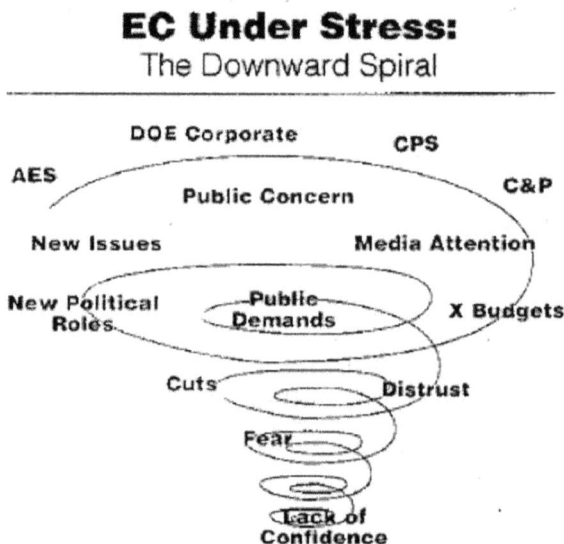

EC Under Stress:
The Downward Spiral

DOE Corporate · CPS

AES

Public Concern

C&P

New Issues · Media Attention

New Political Roles · Public Demands · X Budgets

Cuts · Distrust

Fear

Lack of Confidence

Figure 2.3 EC's downward spiral, 1991[27]

To break out of this decline, the managers identified three challenges which needed tackling: gaining some control over the demands being made on the Department, putting its house in order, and fostering a culture of achievement and of continuous learning. In particular, if it was to meet public expectations, EC needed to start behaving as an integrated Department and not as a "loose association of separate Services".

Before the Department could hope to move forward boldly – as the Green Plan would demand – the Department

---

[26] *Environment Canada in Transition. Report – Year One*, September 20 1991
[27] Ibid.

and its Services had to change from a "holding company" and from being fearful and defensive; innovation and "joint optimization" of service delivery were essential. If the Department was ever to manage the demands upon it satisfactorily, never mind meet the challenges of the Green Plan, then it had to become a cohesive department characterized by trust and self-confidence.[28]

"Transition" was the name given to the two-year, EC renewal process that followed the Montebello meeting. Six areas were identified as needing attention: developing a vision, effectively managing and using EC's knowledge, empowering managers and staff, strengthening operational capabilities, training and developing employees, and working together as one Department and with others through partnerships. The Transition exercise fostered a number of activities within each of the six categories. One of these was an overview of EC's R&D.

In March 1991, the DM (Len Good) asked the Science Advisor (Alex Chisholm) to prepare a strategy for the conduct and management of the Department's R&D. The first phase of work on it was a mainly quantitative profile of EC's R&D and a summary of the key issues concerning the delivery and management of the Department's science.[29] This was presented to the DM and senior management in June. The second phase, the development of the strategy itself, did not commence until December probably because of the effort involved in implementing the Green Plan. Soon after, the task of producing a strategy was overtaken by another request from the DM. In January 1992, he met with the group that had been struck to work on the strategy. He asked for a description of the current portfolio of EC's R&D and of the methods used in the Department to make decisions on R&D. In response to the first charge, a compendium of EC's R&D was assembled after considerable work.[30] The document was organized according to the priority environmental goals in the Green Plan.

The description of how decisions on R&D were made in EC was delivered to the DM and the Departmental management committee in May 1992.[31] The report noted that there was no separate or common planning process in EC for setting R&D priorities. Such planning took

---

[28] Ibid.
[29] *R&D Profile and Issues Concerning the Delivery and Management of Science*, presentation to the Deputy Minister, June 10 1991. EC OSA, box 7
[30] *A Compendium of R&D in Environment Canada*, prepared by the Office of the Science Advisor, October 1992
[31] *A Description of R&D Decision-Making at Environment Canada*, working paper, May 4 1992. EC OSA, box 3

place within each Service according to the regular Departmental decision-making process. The Science Advisor went on to highlight three issues: the Departmental management committee was not involved in a formal manner in strategic planning for R&D; operational planning for R&D was Service-oriented, a situation which did not foster integration across the Services; and, the Departmental decision-making process needed to be more open to science and to scientists.

The Departmental management committee's reaction to the first issue was that the situation was changing and it was now getting involved.[32] The second was seen as difficult, requiring the engagement of senior management. The third concern needed to be handled carefully so as not to cause conflict between scientists and science managers. Nonetheless, it was decided to experiment with openness by inviting a large number of researchers to a Departmental Science Forum being planned for later that year. Its objective was to identify emerging environmental issues from a scientific perspective. Forty of the sixty participants were researchers.[33] The meeting came to a consensus on several strategic directions for EC's R&D. It also strongly recommended the use of science assessments as a way of ensuring the input of science into policy.[34] The Forum was claimed to be the first such meeting in EC with researchers from all parts of the Department. It was certainly not the last time that scientists from across the organization would be enrolled in strategic priority-setting processes.[35]

The move towards an integrated Department, as shown in the Transition exercise, was not the only factor in EC's interest at that time in R&D planning and priority-setting. Concern with S&T policy and with strengthening accountability for resources also contributed. In 1992 and 1993, the National Advisory Board on S&T, which reported to the Prime Minister, discussed the setting of federal S&T priorities with EC and with other science-based departments and agencies. It recommended that they manage their S&T as distinct strategic assets and that a system for S&T priority-setting be established within and among federal organizations.[36]

---

[32] Minutes, Environment Canada Science Committee, June 21 1992. EC S&T Management Committee, box 1

[33] *Report. Environment Canada. Science Forum I,* January 1993. EC OSA, box 3

[34] As a result, EC conducted a thorough science assessment of biodiversity, *Biodiversity in Canada: A science assessment for Environment Canada,* 1994.

[35] Using scientists to brainstorm on priority emerging issues became a fairly common practice. The Federal Innovation Networks of Excellence proposal, the Integration Board and the Beyond the Horizon initiative all used groups of federal scientists for this purpose. For further information on these activities see chapter 7.

[36] *Report of the Committee on Federal Science and Technology Priorities: Spending Smarter,* June 1993. See also its *Phase II* report, February 1994.

The following year, the Auditor General recommended that departments set clearer research goals and priorities.[37] And efforts at that time on a new federal S&T strategy led to recommendations that each science-based department and agency should prepare S&T plans, set clear S&T targets and objectives, and prepare reports on their S&T priorities.[38] However, this external advice, like EC's own reviews, did not cause the Department's planning and reporting to more explicitly address S&T.

Research Agendas

In 1995 EC implemented a new business planning process.[39] Prior to this, planning and reporting had been aligned with the organization of the Department. Now, they were based on the intended results of EC's programs. The three broad outcomes or business lines were: reducing risks to human health and to the environment, providing weather forecasts and warnings and emergency preparedness services, and giving Canadians the tools to build a greener society. All of EC's programs were linked to the three outcomes. The new process was part of an initiative in the federal government to improve reporting to Parliament. EC was among the first departments to adopt the new system. Similar to the focus of the Transition exercise, business lines were seen by senior management as a means of enhancing departmental integration.

The shift to business lines did little to resolve concerns about S&T planning and priority setting. The latter was one of the key S&T management issues set before EC's external R&D Advisory Board at its second meeting in October 1997. The challenges were described as defining R&D priority setting relative to EC's other priority-setting processes; determining whether business planning processes adequately addressed R&D priority setting and whether a formal, department-wide, strategic R&D priority-setting framework was needed; and, enhancing interaction between senior managers and the scientific community in priority setting.[40] The Board decided to form a working group to examine these processes within EC.

---

[37] *Science and Technology – Management of Departmental Science and Technology Activities*, chapter 10 of the 1994 Report of the Auditor General of Canada
[38] *Science and Technology for the New Century: A Federal Strategy*, March 1996
[39] *Environment Canada Business Plan 1995/96-1997/98*, June 1995
[40] Meeting of the R&D Advisory Board, October 14 1997. EC S&T Advisory Board, box 1

For the next year the working group held a number of meetings to discuss priority setting and various other related issues.[41] Among several recommendations it made about business planning, the working group advised the Department to more formally integrate R&D priority setting into the annual business planning cycle. And building on the closely related efforts of the Board's S&T capacity working group, it also recommended that the Department develop science agendas for its business lines.[42]

EC's senior managers, like the Board members, saw S&T planning and reporting as taking place within the framework provided by the business lines. The Department acted on the Board's recommendation about research agendas. By the summer of 2000, five-year research agendas had been completed for the three business lines, at that time known as Nature, Clean Environment, and Weather and Environmental Prediction. Following a request from the DM, the three were synthesized into one Departmental research agenda.[43]

Despite the successful preparation of the research agendas, the Department had difficulties in explicitly incorporating S&T into its planning and reporting. When, for example, the R&D Advisory Board wanted to know the types and amounts of major research and applied scientific activities performed by the business lines, a special study was required.[44] The root of the problem lay in the fact that most of EC's work involved S&T activities, with the intended results being environmental. Reporting on its programs from two different perspectives was unacceptable, seen as an unnecessary duplication of effort. There was no demand for reporting along S&T lines from Treasury Board (which was focused on planning and reporting by results), notwithstanding the ongoing calls for this from external advisory bodies on S&T policy. The result was a Departmental planning and reporting system in which S&T was mostly invisible.

The Science Plan

The creation of the S&T Branch in 2005 provided a focal point for the Department's S&T and the foundation for a more integrated research effort. Its formation along with the reframing of business lines into

---

[41] For example, they discussed external peer review. This led to the adoption by EC of a framework for such reviews – *Framework for External Review of Research and Development in Environment Canada*, S&T Management Committee Report No. 4, February 2000.

[42] The two working groups were merged in 1999.

[43] *Environment Canada Research Agenda*, October 2002

[44] *Research & Development and Related Science Activities in Environment Canada*, Science Policy Branch working paper #7, March 2000

strategic outcome boards created the conditions for EC's first, truly integrated, long-term science strategy. Work on it was triggered by the recommendations of an independent panel on the management of EC's S&T – around research plans, long-range planning and emerging issues – as well as by a request for a strategy from the DM (Samy Watson).[45] Named *Environment Canada's Science Plan*, it was meant to be a strategy to guide the conduct and management of the department's science over the next decade (Figure 2.4).[46]

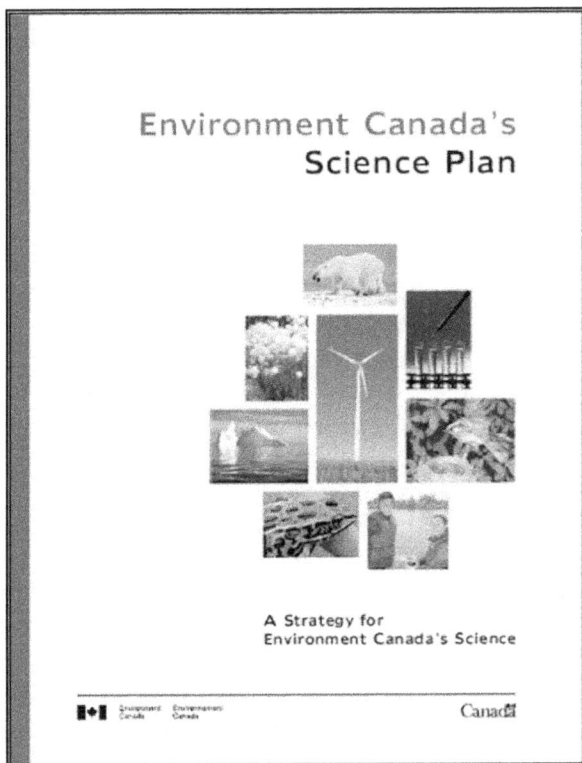

Figure 2.4 Cover of the *Science Plan*

The *Science Plan* was developed relatively quickly through an intense series of activities from January to November of 2006.[47] The first major

---

[45] *Science and Technology Management Review Panel Report*, Environment Canada, November 2004

[46] Technology was not included in the Plan, pending the development and acceptance of a discussion paper on the role of technology in the Department. See chapter 9.

[47] This section is derived from "Building Environment Canada's Science Plan", a presentation I gave at Health Canada on January 16 2007.

step was a week-long workshop in March. Forty-five scientists, seventeen of whom were not from the Department, were brought together to discuss and make recommendations about the long-term science needs and gaps of each of EC's strategic outcomes, as well as potential synergies in that science. Unlike the process for the research agendas in 2000, the long-term science directions were developed for the whole Department, not by business line. The reports from the workshop were reviewed the following month by panels of external experts. Their comments were then used to revise the reports. This material became the basis for the *Plan*'s chapters on the challenges, opportunities and strategic directions for EC's science.

The second major input to the *Science Plan* was a two-day meeting in May of 100 S&T Branch managers and senior scientists. They identified priority science management issues for the Department and suggested actions that could be taken to address them. The results of the meeting fed into a report, which was reviewed by a group composed of science ADMs from selected federal departments and the director of science policy from the US Environmental Protection Agency. This became the source for the *Plan*'s chapter on implementation.

The third step was a series of eight information sessions held across the country in August. Five hundred and thirty five individuals, one hundred and fifty of whom were from outside of EC, attended. The purpose was to share the main ideas in the *Plan* and to obtain feedback, and to make adjustments if necessary.

The final step was a review of the draft *Science Plan* by nine distinguished non-EC scientists and science managers drawn from Canada and other countries. The group met in September. They liked the visionary direction and content of the *Plan*, but suggested that it needed a clearer storyline. The *Plan* was edited with this in mind. The departmental management committee approved the revised version in November.

The core of the *Science Plan* was composed of the chapters on the challenges and opportunities and on the strategic directions for EC's science. The former was organized according to the Department's strategic outcomes, which were the categories for the business lines, or results structure as it was then known. However, the *Plan*'s three strategic directions – developing an integrated environmental monitoring and prediction capability; understanding cumulative risks; and, managing risks, optimizing opportunities and building resilience – cut across those outcomes. The intention was that the three directions would increase integration within the Department by guiding the annual planning process, which was based on the results structure.

The *Science Plan* set a new benchmark for strategic planning of EC's science. However, it did not lead to increased visibility for S&T in the

Department's planning and reporting system. The formal planning and reporting documents of the Department (the *Report on Plans and Priorities* and the *Departmental Performance Report*) remained unchanged. That was not because science had become less important to the Department's work. A special study in 2009 showed that EC's R&D was well aligned with the Department's priority objectives and that there were important hubs of R&D that served more than one objective.[48] The experience with the *Science Plan* showed the barriers posed to S&T planning and reporting by an accountability system focused on high-level environmental results. Since S&T was not formally recognized as a departmental priority, it remained largely indiscernible in planning and reporting documents.[49]

Engaging in Federal Policy for Science

Federal Role in S&T

EC was created at a time of strong federal interest in science policy. Two months after the Department opened its doors, the government established the Ministry of State for Science and Technology (MOSST). Its primary purpose was to formulate and develop policy for science. That mandate, highlighting science as a goal of federal policy, stood in contrast to EC's view of science as a means to an end, creating a fundamental tension in the relationship of the two organizations. Although tiny in comparison to the size of EC, MOSST would have an impact on the Department.[50]

As early as January 1974, EC's departmental management committee was discussing that relationship. Nine areas of concern were identified, including program coordination, the Make-or-Buy policy, science expenditures, university grants, science program evaluation, federal-provincial relations, and international scientific relations.[51] The issue of control over S&T lay behind most of these. EC's view was that its science

---

[48] *Measuring Environment Canada's Research and Development Performance*, 2009
[49] The Commissioner of the Environment and Sustainable Development has also noted the need to integrate the Science Plan with departmental operational planning and with the results management framework. See chapter 2 of his December 2011 report.
[50] MOSST had 41 employees a few months after it opened and a budget of just over $1 million in its first year, while EC was the largest science-based department at that time with over 11,000 employees and a budget of more than $200 million.
[51] LAC Acc. 1993-94/003, box 26

was closely tied to its mandate and should not be subject to separate approval by other departments.[52]

MOSST faced a difficult challenge in its interactions with science-based departments and agencies, which occurred in several ways.[53] It engaged them in senior-level committees. For example, it established a DM-level Advisory Committee on Science Policy, which first met on February 18, 1975.[54] Three years later, a Science ADMs Committee was formed to provide advice to MOSST and to serve as a forum for departments to raise and discuss common issues.[55] The ADM of the meteorological service was EC's representative. Neither committee was very successful. In January 1982, MOSST was proposing that the ADM committee "become a much more vigorous and active body."[56] This was the first attempt, among many in subsequent years, to rejuvenate it. A successor to the ADMs committee still exists today, despite an ongoing general consensus about its ineffectiveness.

A second way in which MOSST engaged science-based departments was through requests for information and for participation in studies. This included annual requests for data on science expenditures, a task later assumed by Statistics Canada. EC responded to these requests, although it never made the collection of information on its science expenditures a regular part of its financial accounting system. MOSST also asked for information on departments' scientific activities. For example, in late 1976 and again in mid-1978, MOSST asked EC about its R&D priorities.[57] The Department produced R&D profiles for the environment and forestry sectors for MOSST in early 1980, as part of the latter's development of plans to deliver on the government's target of increasing R&D expenditures to 1.5% of GNP.[58] Some months later EC was asked to identify areas within the sectors where the government could build up industrial R&D capacity.[59] Soon after, MOSST began a

---

[52] This view was voiced at the Lamontagne Senate Committee in 1976. LAC Acc. 1993-94/003, box 27

[53] For an early critique of its efforts, see Peter Aucoin and Richard French, *Knowledge, Power and Public Policy*, Science Council of Canada Background Study no. 31, 1974.

[54] EC was represented at the meeting by the Science Advisor. LAC Acc. 1993-94/003, box 20

[55] Letter from MOSST to DM of EC, January 27 1982. EC Roots, box 4

[56] *New Role for the Science ADMs Committee*, January 27 1982. EC Roots, box 4

[57] Memo, March 9 1977. LAC Acc. 1993-94/003, box 4. Memo, July 12 1978. LAC Acc. 1993-94/003, box 24

[58] R. G. Lawford, *Research and Development Profile for the Environment Sector*, November 5 1980. See also MOSST's background paper no. 13, *R&D Policies, Planning and Programming*, January 1981.

[59] *A Plan for Federal Encouragement of Industrial and Industry-Oriented R&D in the Environment Sector*, October 15 1980. EC Roots, box 3. See chapter 9 for more on this.

major study of the health of S&T in the federal government. It proposed looking at the factors "which are having a negative impact on the performance, creativity and productivity of government science."[60] This study took place at a time of great concern about federal science, as attested by the Professional Institute of the Public Service of Canada's 1983 discussion paper *A Future for R&D in the Public Service?*

MOSST also involved departments in articulating the role of the federal government in S&T. The subject took up most of the first meeting of the DMs Advisory Committee on Science Policy. EC found MOSST's early attempt at defining the role too narrow, arguing that the purpose of much government science is to provide "information essential to the development of management policies."[61] EC endorsed a revised version of the document, *The Role of the Federal Government in Science and Technology: A Conceptual Framework*, which was approved by Cabinet in May 1975. The *Framework* stated that federal science policy encompassed three policy areas: the support of science, the application of S&T resources, and the use of science in the development of public policy.[62] In deference to departmental authority, the *Framework* went on to affirm that federal science policy must serve the government's goals as represented by the objectives of its departments and agencies, and that these objectives should serve as the primary basis for the development of S&T policies and programs and for the allocation of resources.

In addition, EC helped to develop a MOSST document on the role of federal S&T in the area of natural resources and the environment. The *Framework for Federal S&T Activities in Natural Resource and Environmental Management* was ready in May 1979 and was based on four case studies.[63] It identified seven roles: the development of national inventories and statistics, the resolution of national or regional problems, the management of resources under federal jurisdiction, the fulfillment of international obligations and activities, the enhancement of Canadian industrial S&T capability, the establishment of codes and standards, and the provision of independent advice and certification.

The tension between EC and MOSST did foster the development of EC's policy for science. MOSST's various activities added to the pressure on EC to pay attention to its science, to develop a rationale for those efforts, and to collect information and report on its scientific activities. For its part, EC's goal was to influence federal S&T policy so that it would be supportive of the Department's S&T effort for the

---

[60] Op. cit., footnote 56
[61] Memo, February 12 1975. LAC Acc. 1993-94/003, box 20
[62] LAC Acc. 1993-94/003, box 17
[63] LAC Acc. 1992-93/011, box 56

environment.[64] The Department had some limited success. For example, MOSST policy documents did reflect EC's perspective on the role of government science. And EC did attempt to take advantage of the opportunity to work more closely with MOSST provided by the three years between 1980 and 1983 when its Minister (John Roberts) was also responsible for MOSST.[65] However, EC's engagement with MOSST did not bring about a change in the political agenda for science or increased resources for its S&T. The political priority remained fixed on science for economic development, and within that on increasing industrial R&D capability. EC found its S&T increasingly unconnected with the primary concern of federal science policy, especially when fisheries and forestry – those parts of the Department most directly linked to the economy – were moved out of the organization in 1979 and 1984 respectively. The gulf would continue to grow with the demise of MOSST in the late 1980s.

Federal Strategies for Science

The election of the Brian Mulroney Progressive Conservative government in 1984 led to a short period of renewed federal interest in S&T. A new external advisory body reporting to the Prime Minister, the National Advisory Board on Science and Technology, was created in 1986 and had its first meeting in February 1987. The following month, federal, provincial and territorial ministers signed the country's first *National Science and Technology Policy*. Along with it came a Council of Science and Technology Ministers. At the same time, the government launched *InnovAction*, its strategy for S&T. The primary focus was on industry and to a lesser extent on university science. However, one of the strategy's priorities was to manage federal S&T resources more effectively. To help accomplish this, the government adopted a set of principles, objectives and guidelines for the management of federal S&T, set out in *A Decision Framework for Science and Technology*.

The origin of the *Decision Framework* lay in a request by the Prime Minister in December 1985 for an analysis of federal S&T.[66] The *Decision Framework*'s purpose was to enable Ministers to provide strategic direction to the government's S&T activities through the expenditure management

---

[64] *Federal Science Policy – Impact on DOE and Involvement of the OSA*, [1974]. LAC Acc. 1993-94/003, box 29
[65] John Roberts was Minister for both organizations from March 3 1980 to August 11 1983. For EC's efforts see *Science Policy Strategy Meeting*, April 22 1980. EC Roots, box 10
[66] MOSST, *Annual Report* (1986-87)

process.[67] It was an effort to lessen the tension between government-wide priority setting and decision-making on S&T and departmental responsibilities and accountability.[68] It marked the end of the attempt, through MOSST, to have central agency coordination of federal S&T. In August 1987, the government announced its intention to create a Department of Industry, Science and Technology by combining MOSST and the Department of Regional Industrial Expansion. The new Department opened in 1990. Its focus was on industrial competitiveness and excellence in science.

The *Decision Framework* required departments to produce annual S&T plans, which would be rolled up into an overview for Cabinet. The small energy and science unit within EC's Corporate Planning Group along with the EC Working Group on Science Management pulled these together in 1987 and 1988.[69] They were the first overviews of the Department's S&T activity to be produced for an audience outside EC. However, there proved to be little demand for such documents. The roll-ups for each of the years were prepared but did not reach Cabinet. One further effort at producing an overview of federal S&T was attempted in 1989. This time, departments were not required to produce plans, simply to submit material. A document was published but there was little mention of EC's S&T in it.[70] EC science policy staff felt that the "overall benefit of the exercise is scarcely worth the effort that goes into it."[71] By then EC policy staff were largely consumed with preparations for the government's environmental agenda, the Green Plan. And the new Department of Industry, Science and Technology was preoccupied with organizing itself.

The next Federal strategy for S&T emerged after the election of the Jean Chrétien Liberal Government in the fall of 1993. At the time, Industry, Science and Technology Canada was working on a federal S&T policy document, which EC viewed as unbalanced because it focused on financial returns to R&D and ignored sustainable development.[72] The new government announced a federal S&T review in February 1994, with the intention of putting in place an S&T strategy. A consultation process began, including a discussion paper, an internal review, meetings in

---

[67] *A Decision Framework for Science and Technology: Principles and Guidelines for the Management of the Federal Government's Science and Technology Activities*, March 1987
[68] Memo, October 14 1986. EC Roots, box 3
[69] *Environment Canada Annual Science and Technology Plan 1987*, and *Environment Canada Annual Science and Technology Plan 1988*
[70] *1989-90 Strategic Overview of Science and Technology Activities in the Federal Government*, Industry, Science and Technology Canada, 1990
[71] Memo, November 2 1989. EC OSA, box 1
[72] Minutes, September 15 1993. EC S&T Management Committee, box 1

various communities across Canada, and an assessment by the National Advisory Board on Science and Technology.[73] The internal review took a balanced approach to the government's S&T roles. Four interdepartmental committees were established, dealing with quality of life, the advancement of knowledge, S&T information systems, and job creation and economic growth. EC participated in the committees, which completed their work that fall. In addition, the Auditor General's 1994 report contained several chapters critiquing the management of federal S&T.

The development of the strategy commenced quickly and was actively pursued in 1994. However it soon ran into difficulties. The government's Program Review, which was being undertaken at the same time, became the primary focus of attention. Additionally, there were internal problems in Industry, Science and Technology Canada. The government released the strategy in March 1996. *Science and Technology for the New Century: A Federal Strategy* emphasized the need to pursue economic growth, quality of life and the advancement of knowledge in "an integrated and mutually reinforcing way." It also sought a better governance system for federal S&T and set out seven principles to improve the management of the federal S&T effort.

EC participated in the release of *Science and Technology for the New Century* and its follow up. The strategy was accompanied by departmental S&T action plans. EC was well positioned to produce its contribution, *Environment Canada's Science and Technology: Leading to Solutions*. This was because the Department had already developed its own plan for managing its S&T in order to deal with the various reviews underway in 1994.[74] The follow-up to the strategy required EC and other science-based departments and agencies to provide information on what they were doing to implement the strategy. Six annual reports on the strategy were issued by Industry Canada (as it was then named) until 2003. EC contributed information to all of them.

Despite EC's involvement in the preparation and reporting on *Science and Technology for the New Century*, the content of the strategy had little direct impact on the Department. The biggest consequence was for EC's policy for science. The strategy caused the establishment of an external R&D Advisory Board to EC's DM.[75] And it gave birth two years later to the Council of Science and Technology Advisors, an external advisory

---

[73] The discussion paper was titled *Building a Federal Science and Technology Strategy*. The Advisory Board's report was *Healthy, Wealthy and Wise: A Framework for an Integrated Federal Science and Technology Strategy*, April 1995

[74] *Action Plan for Managing Science and Technology at Environment Canada*, November 1994

[75] See chapter 1 for more on the Board, and other chapters for aspects of its work.

board to the Minister of Industry, which was devoted to S&T performed by federal departments. EC's policy for science capacity expended a great deal of effort in responding to and working with these advisory bodies. The engagement proved to be worthwhile as both groups produced reports and recommendations of value to the Department.

The return of the Conservatives to power under Stephen Harper in early 2006 led to another federal S&T strategy. The government's budget speech in April announced that the ministers of industry and of finance would be responsible for carrying out an S&T review and producing a federal S&T strategy. Unlike the process a decade earlier, there were limited consultations – mainly with Industry Canada's contacts – and federal departments were not engaged, only given general updates. The strategy, *Mobilizing Science and Technology to Canada's Advantage*, was released in May 2007.

The strategy was a collection of specific initiatives rather than a comprehensive plan. Its focus was on the economy. The message from the Minister of Industry in the document stated that the "most important role of the Government of Canada is to ensure a competitive marketplace and foster an investment climate that encourages the private sector to innovate." Pressure from science-based departments to get the strategy to pay attention to federal S&T and its various roles led to only a brief mention. A short section towards the end of the strategy, "Exploring New Approaches to Federally Performed Science and Technology", committed to focusing the government's S&T on where it "is best able to deliver benefits to Canadians" and to enhancing collaboration within the federal S&T community. That section also repeated the government's commitment to transfer management of some non-regulatory federal laboratories to universities.[76]

The interdepartmental ADM Committee on S&T was given responsibility for overseeing implementation of the strategy's 36 commitments. It decided to concentrate on those that had few if any associated action items. Among them were the general commitments about S&T in the government. The ADM Committee established several interdepartmental working groups in the summer of 2007 to come up with ways in which the government might move forward with the commitments. EC participated in the groups, chairing the one on directing resources to priority areas. The latter met a dozen times over the next year and a half. Its report, *Directing Federal Resources to Priorities* (February 2009), identified the main challenge to be the need to increase coordination across the government. Among its recommendations was a DMs' committee to strategically manage whole-of-government S&T

---

[76] See chapter 8.

investments in priority areas. The working group's report met with a cool reception at the ADM Committee. However, several months later, a DMs' committee was organized to look at federal S&T. While likely not a direct result of the working group's report, the creation of the committee was an acknowledgement of the need for a better governance system for federal S&T.

EC devoted a large share of its policy for science capacity to federal S&T strategies. It provided input, implemented relevant parts of the strategies, and reported on progress. However, these strategies came to focus almost exclusively on federal investments in industrial and academic R&D from the perspective of economic competitiveness. Their content reflected the abandonment of federal efforts to coordinate its own S&T, as signalled in the closure of MOSST and the activities of its successors. Without much orientation to economic development, EC was little impacted by the federal S&T strategies. Despite its involvement in them, EC was unable to influence the strategies to address the goals and needs of its S&T. Political priorities would need to shift, if federal S&T policy was to support environmental S&T.

3

Pursuing Strategies for Northern and International S&T

Two subsets of EC's scientific activity – one devoted to the North, the other to international efforts – have regularly generated calls for departmental strategies. Like the appeals for a departmental science strategy, the impetus has come both from within the Department and from other parts of the federal government. Much policy work has resulted, but not strategies. The complex nature of northern and international S&T is partially responsible for the lack of success. But the major factor has been the weak demand for such strategies from either EC or from the federal government. This chapter reviews the sporadic policy efforts to develop strategies for EC's northern and international S&T.

Northern S&T Strategies

Over 40% of Canada is in the Arctic. The country is the second largest Arctic nation. It should not be surprising then that the North has been an object of concern, study and action by EC from its beginning. Indeed, many of the groups that were brought together to constitute the Department already had a long history of working in the North. EC has always been one of the biggest performers of northern S&T both in the federal government and in Canada. This stems from its mandated responsibilities, among them weather forecasting, sea ice, migratory birds, environmental regulations, and water quantity and quality.

The Department's scientific effort in the North has faced many challenges. Despite its numerous activities there, the proportion of EC's effort devoted to the region has never been large. The Department's focus has been on the southern parts of the country where the large majority of Canadians live. In addition, the significant differences in purpose, clientele and scientific discipline among the Department's components translated into a stove-piped approach to northern S&T. The sparse population, lack of infrastructure and harsh conditions in the North meant very high operating costs for science. However, resources for northern S&T have not usually matched the requirements for dealing with environmental issues there.

These challenges have both driven and frustrated many attempts in the Department to develop its own strategy for the North and to coordinate its activities there. EC's lack of success was also due to the absence of an enabling framework at the federal level. Much of the Department's policy work on northern S&T was in support of interdepartmental efforts to develop a federal northern S&T strategy or engage in international polar initiatives. These efforts were led by the Department of Indian and Northern Affairs, whose responsibilities in the North included the coordination of federal activities there. But that Department, like others in similar situations, struggled with balancing the advancement of its own agenda and interests with attending to the needs and well being of other departments.

The Challenge of an EC Northern S&T Strategy

One of the earliest policy issues for EC science was the building of new research institutes. At the same time, the Department was thinking about Northern environmental issues and what it might do about them.[1] These two interests came together, giving rise to a proposal for an EC Arctic Environmental Institute. When the departmental management committee met in May 1972 to consider the proposal, it opted instead to set up a Northern Program Advisory Group, to be chaired by the newly appointed Ken Hare.[2] The Group met the following month and decided to pull together an overview of EC's current activities in the North. The result, published as *Northern Programs of Environment Canada*, showed that the Department was spending $13.4 million on the North, about 5% of its total budget.[3] The Atmospheric Environment Service accounted for the largest share, at $7 million; with $3 million spent by Water Management, $1.9 million by Lands, Forests & Wildlife, $1.3 million by Fisheries, and $0.2 million by Environmental Protection. The Group presented its final report to the departmental management committee in November.[4] It noted that there was a "clear need for the Department to establish a policy to guide its work in the North." The Group recommended that EC's northern activities be coordinated at the field level by regional organizations, at the level of research by the Research Coordination Directorate, and interdepartmentally by the Advisory Committee on Northern Development.

---

[1] *Opportunities for the DOE in the North* was one of the proposals of the Main Chances exercise. The latter was an effort to develop an agenda of environmental priorities requiring concerted effort across the Department. LAC Acc. 1993-94/004, box 2

[2] Meeting, May 3 1972. LAC Acc. 1991-92/017, box 3

[3] *Northern Programs of Environment Canada*, August 1 1972

[4] Meeting, November 30 1972. LAC Acc. 1991-92/017, box 5

Looking to the Advisory Committee on Northern Development for interdepartmental coordination was an obvious choice. A DM-level committee, it had been established in 1948 to provide advice on policy relating to civilian and military activities in the North and to coordinate federal activities there. It had many subcommittees, including one on S&T. In 1970 the Advisory Committee conducted a review of federal S&T in the North. It concluded that guidelines and priorities were needed which went beyond departmental goals and were aligned with national objectives and plans for the North.[5] The Advisory Committee decided to organize a seminar that would bring together scientists and other interested parties from universities, the private sector and government to discuss the guidelines and priorities. One of the six main areas to be examined was the natural environment. While still at the University of Toronto, Hare had been asked to write the background paper for that area. He sought and obtained the input of EC expertise in its preparation.[6] Upon joining the Department, he also worked with the Northern Program Advisory Group to develop a departmental position for the seminar.

The seminar was held in October 1972. However, it did not lead to much action, either by the Advisory Committee's S&T subcommittee or by the Department of Indian and Northern Affairs. EC was unhappy about the lack of progress. The Department was a major player in Northern S&T. An inventory of federal science activities in the North during 1975-76 showed that EC was the most active federal department, accounting for $15 million out of a total of $42 million in expenditures.[7] Fred Roots, an expert on the North, felt that the Advisory Committee had ceased to be a "vigorous and genuine" interdepartmental committee, becoming simply a "sounding board" for Indian & Northern Affairs.[8] The situation was further complicated by jurisdictional tensions. Some parts of Indian & Northern Affairs were not pleased by the decision that MOSST should review northern science policies and priorities to ensure that they were assessed in the same way as other science policies and were consistent with national science policy frameworks.[9] In 1977 MOSST

---

[5] *Science and the North: A Seminar on Guidelines for Scientific Activities in Northern Canada 1972*, 1973
[6] *The Natural Environment of the Canadian North*, EC, August 31 1972
[7] *Inventory of Federal Northern Science Projects, 1975-76*, MOSST, 1977. An earlier survey, *Federal Government Research and Scientific Activities in the Arctic and Sub-Arctic, 1971-72*, had been carried out before the creation of EC. An annual listing of EC's activities in the North at that time can be found in the publication *Government Activities in the North*.
[8] Memo, June 29 1976. LAC Acc. 1993-94/003, box 6
[9] Memo, February 24 1975. LAC Acc. 1993-94/003, box 20

was recommending development of a framework for the planning, funding and assessment of northern S&T.[10]

By the late 1970s, there was a renewed interest in EC in pulling together its northern S&T programs. This reflected an ongoing Canadian interest in the North. The Science Council, for example, had issued a number of policy studies at that time on S&T in and for the North.[11] EC's Canadian Environmental Advisory Council had also looked at the status of environmental research in the North in 1977, finding it "seriously inadequate and incoherent" and calling for the development of a long-term plan.[12] In December 1979, the departmental management committee discussed EC's policies and programs for the North.[13] The DM directed the Corporate Planning Group in August 1980 to prepare a departmental policy on the northern environment, a statement on its responsibilities and roles, and an action plan to guide EC's programs in the North.[14] This led to the 1983 discussion paper *Environment Canada and the North: The Perceptions, Roles and Policies of the Department of the Environment Regarding Development North of 60°*. It stated that the Department would undertake research to promote environmentally sound technology and safe operations in northern development.

EC's policy work on northern S&T in the late 1980s and into the 1990s mostly consisted of meetings, sponsored by the Department's Prairie and Northern Region, to talk about the need for a more coordinated and strategic approach. In June 1989, a Northern Science Seminar was held in Inuvik. It provided an overview of EC's scientific activities, assessed requirements, and considered strategies for long-term science programs. Five years later, in March 1994, an Arctic strategy meeting took place in Edmonton. It also reviewed the Department's science programs, calling for greater coordination among the various groups involved. A similar gathering was held in February 1997, again in Edmonton. Like the earlier meetings, this one looked at EC's activities, discussed the need for improved management, coordination and communications, and called for a northern policy to guide that work.

While developing a strategy for EC's northern activities proved challenging, it was not a roadblock to doing northern S&T. The Department's Services developed their own programs. For example, the large Northern River Basins Study was started in 1991, the Arctic

---

[10] *A Commentary on the Inventory of Federal Northern Science Projects, 1975-76*, 1977. Cited in *Canada and Polar Science*, March 1987.

[11] See, for example, its report *Northward Looking: A Strategy and a Science Policy for Northern Development*, 1977.

[12] Letter to Minister, February 8 1977. LAC Acc. 1992-93/011, box 45

[13] Meeting, December 13 1979. LAC Acc. 1991-92/017, box 33

[14] Meeting, February 17 1982. LAC Acc. 1991-92/017, box 38

Stratospheric Ozone Observatory began operations in Eureka in 1993, and the Northern Ecosystems Initiative commenced in 1998.[15] These programs reinforced EC's position as one of the largest performers of S&T in the North. However, the development of policy for that region was mostly left to the leadership of Indian Affairs and Northern Development, the department with the mandate for coordinating federal involvement in the North.

Arctic Environmental Strategy

The Arctic Environmental Strategy was launched in April 1991 by the ministers of the departments of Indian Affairs and Northern Development and of the Environment. It had been announced several months earlier in the Green Plan, one of whose key goals was to preserve the integrity of Canada's North.[16] Like the other components of the Green Plan, the Strategy was spread over six years. Almost its entire $100 million budget was managed by Indian Affairs and Northern Development. The Strategy consisted of four action programs: contaminants ($35 million), waste clean up ($30 million), water management ($25 million) and environment and economy integration ($10 million). The contaminants and water programs were the main scientific efforts. The former sought to understand the risks to the Arctic from the long-range transport of persistent contaminants; the latter to develop and implement water quality and quantity monitoring networks and laboratory facilities in the North.[17]

The Northern Contaminants Program was the only component of the Strategy to be extended after 1997. The investment had assisted in Canada's response to that class of pollutants and placed the country in an international leadership role.[18] EC had received substantial funding from it. When Indian Affairs and Northern Development sought renewal of the Program in 1997, Treasury Board Secretariat (TBS) indicated it would be willing to provide about half of the required budget.[19] The remainder would have to come from participating departments. In addition, TBS wanted the Program to be oriented to human health. The financial condition became a source of friction between EC and Indian Affairs and

---

[15] The Northern Ecosystems Initiative continued until 2009. It consisted of 152 projects with a total investment of about $38 million. *Building Capacity and Delivering Results: Environment Canada's Contributions to the Northern Environmental Agenda 2003-2008*, 2009

[16] The goal had been included in the April 1989 Speech from the Throne.

[17] *The Arctic Environmental Strategy: Five Years of Progress*, 1996

[18] *Evaluation of the Arctic Environmental Strategy. Final Report*, Department of Indian Affairs and Northern Development, 1996

[19] The budget was $27 million over five years.

Northern Development. The distributed nature of EC's science activities meant that the Department had difficulties in determining the level of funding it would contribute and in accepting that the Program's management might then reallocate that funding from one of EC's Services to another. These problems were resolved, and TBS funding was in place by the end of 1997. The Northern Contaminants Program was still operating in 2010.

Northern S&T Strategy

The renewal of the Northern Contaminants Program was one element of a broader plan for Northern S&T. Indian Affairs and Northern Development wanted to develop a strategy for federal S&T in the North that would replace the Arctic Environmental Strategy when it ended in 1997. To assist in the preparation of a common strategy, it created the interdepartmental ADMs Committee on Northern S&T in May 1996. The Committee's objectives were "to develop a strategy to allow the federal government to meet its northern science needs, encourage future collaborations and partnerships, and develop a cohesive northern science strategy."[20] EC's member was the ADM of the Environmental Conservation Service, the lead for science policy in EC.

A discussion paper on the changing needs for science in the North was commissioned and circulated in September.[21] A rather lengthy draft strategy (over 100 pages) was then prepared. The ADMs considered it at their meeting in April 1997. They agreed with the draft's proposed principles and objectives, but felt that its commitments to action were weak. In particular, the ADMs were unsure about how the strategy's joint initiatives would be collectively managed. EC suggested using the *Memorandum of Understanding on Science and Technology for Sustainable Development*, but this was not widely accepted.[22]

To help strengthen the commitments to coordinated action, the ADMs Committee sponsored an interdepartmental workshop in February 1998 aimed at engaging practitioners of northern S&T. In total, there were 72 participants from 14 departments and agencies. EC's Services and several of its Regions with northern functions were represented. Participants were concerned about the diminished capacity of their departments, after Program Review, to provide the S&T needed to deal with northern issues. In their view, this caused inflexibility in

---

[20] *Northern Science & Technology: The Role of Environment Canada*, discussion paper, 1996.
[21] *The Changing Needs for Science in Northern Canada and the Arctic: A Discussion Paper*, prepared by E. F. Roots, September 1996
[22] For more on the *Memorandum*, see chapter 7.

existing programs that made cooperation harder. They echoed previous calls for a national direction and vision for northern S&T, overarching coordination mechanisms, delivery on Canada's international commitments in the North, and improved northern science capacity.[23] The issue that emerged as a top priority at the workshop was the decline in federal support for northern S&T facilities and logistical services, in particular the budget of the Polar Continental Shelf Project.[24]

There was little progress on the draft Northern S&T Strategy in the year following the workshop. The ADMs Committee met only once in 1998, in September, when it asked for options for taking the Strategy to Cabinet and for increasing resources for northern S&T infrastructure and logistics. Indian Affairs and Northern Development appeared to have lost interest in the Strategy. By early 1999, EC, Natural Resources Canada and Fisheries & Oceans decided to move forward on their own. They proposed to revitalize the Committee by having its chair rotate among the members, holding more frequent meetings, using the interdepartmental working group as its secretariat, and developing a pragmatic, deliverable-oriented work plan. In that vein, they wanted to use the material developed to that point for the Strategy, to produce a framework and action plan. Organizational changes at Indian Affairs and Northern Development led to the working responsibility for S&T being reassigned. That department then also became engaged in the revitalization effort.

Northern Science and Technology in Canada: Federal Framework and Research Plan, April 1, 2000 – March 31, 2002 consisted of guiding principles and objectives, as well as an outline of major federal northern research activities planned for the next two years. It was approved by the ADMs Committee in March 2000 and released by the Minister of Indian and Northern Affairs in August. With the completion of the Framework, Natural Resources Canada took over as chair of the ADMs Committee.

The plan for moving forward after the Framework was to make a presentation to Cabinet on the needs and priorities of northern S&T, followed by a request for funding. Work on it began in the spring of 2000. The intention was that the Minister of Natural Resources would make the presentation in the fall. However this was delayed due to the federal election in November. It was not until June of 2001 that the Minister was able to deliver his pitch. He laid out the need to strengthen Canada's ability – through science capacity, logistics and infrastructure –

---

[23] Documents from the workshop held at Strathmere on February 17-18 1998. Science Policy Division files

[24] The Polar Continental Shelf Project was created in 1958. EC depended on the project for support in undertaking northern S&T. From 1993 to 1997, EC used $5 million worth of support, $2.3 million of which was cost recovered from the Department by the Project (which was part of Natural Resources Canada).

to carry out and coordinate S&T research on northern issues in partnership with northerners. His request to come back with a memorandum to cabinet was supported.

Work on such a memorandum had commenced a year earlier, at the same time as the development of the presentation. It was meant to establish a long-term strategy for Northern S&T. However, it faced a number of difficulties. One challenge was to position it among competing requests either on the North (which included S&T elements) or on S&T (with northern components). In addition, there was a question as to whether the memorandum should simply propose a single program to strengthen the foundation of the northern S&T system or include departmental requests for their programs. A change in ADMs at Natural Resources Canada in the fall of 2000 also slowed progress. The pace of development quickened after the June 2001 presentation, but then it stalled. The memorandum had been seeking $350 million over five years. After the al-Qaeda attack on the US on September 11, 2001 and the ensuing economic downturn, the request was no longer seen as viable. Finally, the initiative lost its champion in early 2002 when the minister at Natural Resources Canada (Ralph Goodale) was shuffled to another department.

Despite the lack of a federal strategy, EC continued to actively engage in S&T in the North. The Department was still the largest federal S&T performer in the North, spending an estimated $25 million on northern S&T in 2000 and about $42 million in 2005.[25] This ranking also held when judged by the output of peer-reviewed journal articles (Figure 3.1).

EC's northern S&T effort was delivered through a wide range of programs managed under the Department's various business lines (Figure 3.2). From time to time EC attempted to coordinate this activity by trying to develop a strategy – e.g., the May 2004 draft *Northern Strategic Framework* – and by describing the northern issues it dealt with and the work it carried out.[26] However, the Department's attempts to develop a guiding policy or coordinating mechanism for its northern activities proved no more successful than the interdepartmental ones.

---

[25] Figures are taken from *Northern Science and Technology in Canada: Federal Framework and Research Plan, April 1, 2000 – March 31, 2002*, 2000 and from *Northern & Arctic Meteorological Programs*, 2005. Expenditures for 2005 include about $11 million for the International Polar Year.

[26] *Linking Environment Canada to Canada's North: A Resource Book*, prepared by Gartner Lee Ltd., February 2005

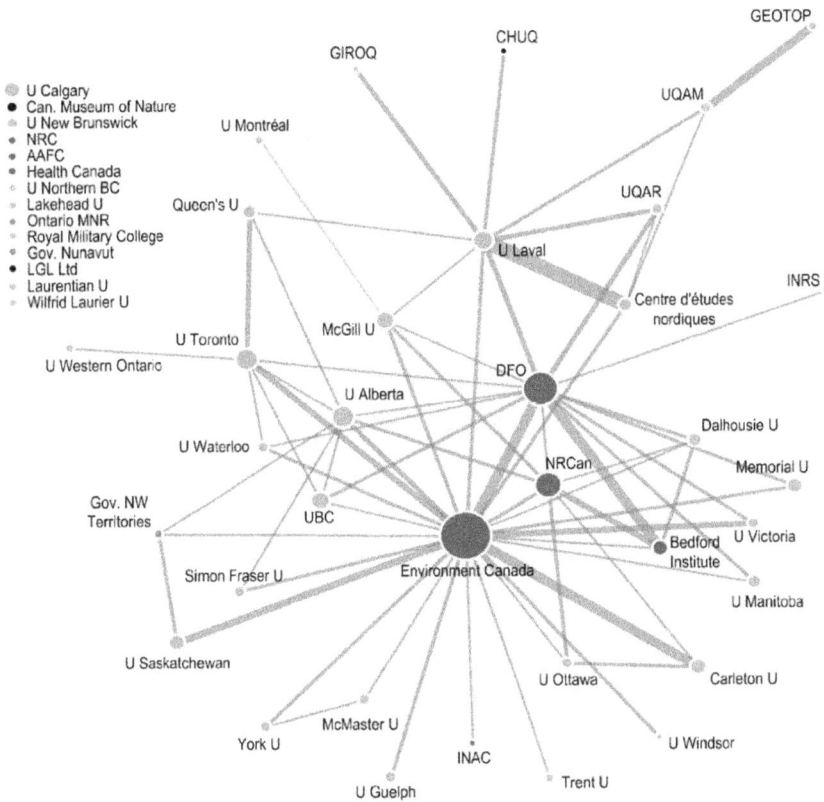

Figure 3.1 Relative number of scientific articles (circles) and of collaborations (lines) in Arctic research in Canada by institution, 1996-2007.[27]

---

[27] *Arctic Research in Canada: A Bibliometric Analysis*, prepared by Science-Metrix for Indian and Northern Affairs Canada, 2009. This figure was constructed using the Scopus database.

	Weather & Env Predictions	Clean Environment	Nature	Management and Policy
Regional Programs	Water & Weather Monitoring Networks	Environment Enforcement	Protected Areas	Program
National Programs	Weather Forecasting Ice Surveys	Technology Development	Legislation - Species at Risk Act	Aboriginal Strategy
Research	Climate Change Research	Kimberlite Toxicity - Diamond Mines	Contaminants in Migratory Birds	
International Research	Global Atmospheric Watch lab (Alert)	Oil spill Response in Cold Climates	Polar Bear Research	
International Policy	Arctic Monitoring Assessment Program	Protection Arctic Marine Environ. /Emergency Prevention, Preparedness and Response	Conservation of Arctic Flora & Fauna / Sustainable Development	International Affairs

**Figure 3.2 Key northern S&T areas by
EC Business Line (columns) and type of program.**[28]

Northern S&T Initiatives

Following the demise of the Natural Resources Canada-led effort to develop a northern S&T strategy and obtain funding for it, the ADMs Committee met sporadically to consider opportunities for federal S&T in the North. For example, the Committee explored advancing its agenda through the Federal Innovation Networks of Excellence (FINE) and later through the work of the Integration Board.[29] In June 2002, it created a working group to update the 2000 Framework, releasing the new version in 2004.[30]

---

[28] EC briefing note, May 19, 2000. Science Policy Division files
[29] For a discussion of FINE and of the Integration Board see chapter 7.
[30] *Northern Science and Technology in Canada: Federal Activity Report, April 1, 2004 – March 31, 2006*

Encouragingly, some new federal funding began to appear. The federal budget in 2003 announced $6 million over two years for the Polar Continental Shelf Project. This was soon followed by some other funding for northern research, including money to retrofit the Coast Guard icebreaker *Amundsen* and granting council support for a multi-year study of melting ice in the Arctic Ocean. However, it was an international scientific initiative – the International Polar Year in 2007-2008 – that would elicit the largest single investment in Arctic research made in Canada up to that time.[31]

The fourth International Polar Year had been proposed by international scientific bodies, organized through the International Council for Science and the World Meteorological Organization. The ADMs Committee on Northern S&T had set up a working group in 2004 to explore options for federal involvement in the Year, and had provided some financial and in-kind support for a national secretariat. Indian and Northern Affairs led the development of a memorandum to cabinet that was supported by six departments, including EC. The result was an announcement in September 2005 of $150 million for Canadian participation in the Year. The amount was to be spent over the six years from 2006 to 2012. The two priority areas for the science program component were climate change impacts and adaptation and the health and well being of northern communities.

EC was much involved in the governance of the International Polar Year. It supported the Year through its membership in the World Meteorological Organization. One of its science managers was a member of the joint International Council for Science and World Meteorological Organization committee for the Year, another sat on the Canadian National Steering Committee, and the ADM of the S&T Branch participated in a new interdepartmental ADMs committee on the Year.

The Department was also one of the most actively involved Canadian research organizations in the International Polar Year. It received over $14 million for research, weather and other logistic services, upgrades to northern research infrastructure, and outreach. Out of a total of 45 peer-reviewed projects funded by the government's International Polar Year program, EC scientists were the principal investigators for five, received funding under another five, and provided their expertise or in-kind support in nine other projects.[32]

In the summer of 2007, the ADMs Committee on Northern S&T was presented with another opportunity to shape Canadian science in the Arctic. The government indicated its interest in promoting an Arctic

---

[31] *Environment Canada's International Polar Year Achievements,* June 2010
[32] Ibid.

agenda, one in which science would have a prominent part. Indian and Northern Affairs began to coordinate the development of the agenda, engaging the ADMs Committee. The focal point of the science component was an Arctic research station, and associated with it were an integrated monitoring program, a research program and a network. The government announced its intention, in the October Speech from the Throne, to build a world-class, international arctic research station.

EC was involved in the discussions. It had a special interest in the integrated monitoring program, which directly aligned with one of the strategic directions in its recent Science Plan. Due to the lack of a departmental coordinating group for northern issues, the Science Policy Division worked to keep the various northern groups informed and engaged in providing input to the interdepartmental work. EC scientists and users of that work were brought together in a workshop in March 2008. Out of it came an EC position paper identifying research and monitoring priorities, as input for the Arctic Research Station.[33]

The development of the Arctic Research Station initiative progressed, although more slowly than had originally been anticipated. The January 2009 federal budget announced $2 million for a feasibility study for the Station. Budget 2010 provided $18 million over five years to start the pre-construction design phase. In August of that year, the Prime Minister announced that the Station would be located in Cambridge Bay. EC has continued to be involved in its planning and associated science programs, seeking to leverage Northern S&T capacity in fulfillment of the Department's objectives.

International S&T Strategies

From its beginning, EC has been involved in international activities. In some cases this stemmed from the mandate of the Department, for example from obligations concerning migratory birds or the International Joint Commission. In others, it was due to Canada's membership in intergovernmental bodies or to the government's commitments to foreign aid. EC's engagement in international cooperation also grew substantially. This was largely a result of the increasing number of environmental issues and to the fact that many of them transcended national boundaries.

A major portion of EC's international work has depended on the Department's S&T capacity. The latter has contributed to all types of

---

[33] *An Integrated Monitoring and Research Program: Canada's Global Advantage in Arctic Science*, EC, April 2008

international collaboration – from formal multilateral agreements to informal joint activities by researchers – and to varied Departmental objectives: environmental, economic, foreign assistance or scientific. The Department's S&T effort has itself become more international. Over 40% of the research articles produced by EC scientists in the period 2003-07 were co-authored with scientists from other countries, compared to 13% in 1980-84.[34]

While EC has been active internationally, the effort has taken place in the absence of a Departmental strategy on international relations. Despite some calls for one over the years, EC never developed such a strategy. This was probably due to the tendency to view international activity as simply the means to domestic environmental goals and therefore not in need of its own strategy. In addition, the great diversity in the requirements of different environmental issues, in the operational needs and objectives of the divisions of the Department and in the types of international engagement, made the development of an EC international strategy very daunting.

EC's policy work on international S&T has also never been a priority. Not only was there little demand for it from within the Department, but federal S&T policy also rarely created any pull either. Federal work on international S&T was sporadic and usually at the periphery of S&T policy priorities. Likely due to similar factors as those affecting the development of an EC strategy, there has not been a federal strategy for international S&T. Given the Departmental and federal context, the paucity of EC policy work on international S&T should not be surprising.

EC Policy for International Activity

The year before the establishment of EC, the Department of External Affairs created an interdepartmental committee on external relations. This was done to improve discussion and coordination of the international activity in which almost all departments were engaged.[35] The new Department of the Environment was no exception. It was involved in many international initiatives. Its organization included an Intergovernmental Affairs Directorate, part of the same Service as Research Coordination. An overview of EC's international work in 1977 took up about 100 pages, and included many bilateral environmental S&T

---

[34] *A Bibliometric Analysis of R&D at Environment Canada.* PowerPoint presentation, Science-Metrix, 2009. *25 Years of Canadian Environmental Research: A Scientometric Analysis (1980-2004)*, Science-Metrix, 2006
[35] Jocelyn M. Ghent, *Canadian Government Participation in International Science and Technology*, Background Study no. 44, Science Council of Canada, 1979

agreements and multilateral S&T organizations.[36] Despite the activity, there does not appear to have been any effort to develop a Departmental strategy for international undertakings at that time.

The situation changed in the early 1980s. Intergovernmental Affairs conducted or commissioned several studies of EC's international activities. One noted that the Department had no international objectives.[37] It proposed five categories of objectives: political, economic, scientific, environmental and cultural. Another inventoried EC's international relations for the first time in many years.[38] A third recommended the Department develop an international policy.[39] All the reports reflected an awareness of the growing importance of the international environmental scene and a desire to define the Department's interests in that work. In 1984, *Proposals for an International Policy: Department of the Environment* was issued. It urged that EC develop an international policy based on three objectives: use of the international system to better fulfill the Department's mandate, identification and management of external environmental problems, and contribution to international economic relations and to foreign aid. Regardless of the interest, no departmental policy for its international work followed. However, two more extensive inventories were prepared, one in April 1985 and the other in November 1986.[40]

The 1990 Green Plan included an international theme. One of its priority objectives was to accelerate global cooperation, understanding and progress on environmental issues. The Department pursued this through increased funding for key international environmental institutions. The magnitude of the Department's international effort was again revealed in another *Inventory of International Affairs*, published in June 1992. It catalogued the Department's involvement in multilateral and bilateral agreements, conventions and protocols. Although the *Inventory* was over 100 pages, it did not capture all of EC's international activities. Foreign aid activities and international research projects, which had been listed in the previous inventories, were not included.

---

[36] *An Overview of International Intergovernmental Environmental Relations*, Liaison & Coordination Directorate, EC, April 28 1977

[37] Richard Kinley, *The International Objectives of Environment Canada: A Synthesis Report prepared for the RIME Study*, Intergovernmental Affairs Directorate, March 7 1983

[38] *Provisional Inventory of the Department of the Environment's International Relations*, Intergovernmental Affairs Directorate, September 1983

[39] Paul Painchaud, *Proposals for a Department of the Environment's International Policy*, Intergovernmental Affairs Directorate, September 1983

[40] *Inventory of International Activities*, Intergovernmental Affairs Directorate, April 1985, *Inventory of International Activities*, External Relations Directorate, November 1986

The appointment of a Commissioner of the Environment and Sustainable Development in the Office of the Auditor General in 1996 stimulated some action within EC to better coordinate its international activity. The Commissioner's first report, in March 1997, announced that his office would conduct a study of Canada's international environmental obligations and provide an overview of the extent to which Canada was meeting its commitments. A year later, the Commissioner reported back that the government did not systematically track the implementation of those commitments.

In April 1998, EC's management committee launched a review of its international work. The resulting report made a number of suggestions about the management of international activities, including the need to strengthen the tracking of international activities and of reporting on them, to create a departmental process to provide oversight and cohesion, and to develop an international strategy.[41] A coordinating committee, chaired by the DG of International Relations, was established that same year. A *Compendium of International Environmental Agreements*, listing 56 legally binding international environmental agreements and organized by priority environmental issue, soon followed. However, the Department still did not develop an international strategy.

The reports of the Commissioner of the Environment and Sustainable Development continued to put pressure on the Department to improve the management of its international activities. The March 2008 report, for example, criticized departments, including EC, for not providing Parliament and Canadians with complete information on the objectives, means and results of their international agreements.[42] In that year, EC developed an International Engagement Framework to help guide the Department's involvement in international activities and keep their focus on achieving environmental results. Although the first such policy for the Department, the Framework was quite general. For example, it set out high environmental risk, opportunity to make a difference and legitimate federal role as the criteria for deciding on engagement. Some effort was put into developing an implementation plan for the Framework, but that was soon abandoned for other activities.

The International Engagement Framework stimulated thinking about international policies among some parts of the Department's management system. The S&T Branch began to explore an international

---

[41] *Review of Environment Canada's International Activities. Final Report*, International Relations Directorate, July 1998
[42] "Management Tools and Government Commitments – International Environmental Agreements," chapter 8 of the *2008 March Status Report of the Commissioner of the Environment and Sustainable Development*

S&T strategy in 2009. However, this did not become a priority. The lack of a department-wide strategy for international activity created a deterrent to developing an international S&T strategy.

Federal Strategy for International S&T

The period during which EC was created was a time of "great expansion in Canada's international S&T relationships".[43] The proliferation of S&T agreements produced a need for their greater coordination and management.[44] When MOSST set up the DM-level Advisory Committee on Science Policy in 1975, External Affairs proposed a subcommittee on international S&T.[45] It was intended to be a forum to exchange views, coordinate participation and recommend policies for international S&T. The first meeting of the Interdepartmental Committee on International S&T Relations was held on March 10, 1975. EC's representative was its Science Advisor, Fred Roots. Three years later, Roots asked the international group in EC to assume departmental representation due to staff cuts in the Office of the Science Advisor.[46]

In 1982, the government announced a strategy on international S&T collaboration.[47] It aimed to increase such collaboration in order to better achieve Canada's economic and foreign policy goals. Soon after, EC's International Affairs Directorate prepared a report on the Department's international S&T.[48] Reflecting the government's international priorities, it emphasized EC's contribution through promoting environmental technologies abroad.

There was little EC policy work on international S&T for the next decade. Then, a few initiatives, stemming from S&T advisory councils and from federal S&T strategies, prompted some effort. In May 1994, the Prime Minister's National Advisory Board on S&T reported on Canada's approach to international S&T.[49] But it had little impact on EC because its focus, like much other federal S&T policy at the time, was on assisting small and medium-sized firms. The accent on the economy was repeated

---

[43] Op. cit., footnote 35

[44] See, for example, report no. 20 of the Science Council of Canada, *Canada, Science and International Affairs*, April 1973.

[45] Memo, February 7 1975. LAC Acc. 1993-94/003, box 6

[46] Memo, April 1978. LAC Acc. 1993-94/003, box 16

[47] Paul Dufour, "Taking the (right?) fork in the road: Canada's two-track approach to domestic and international science and technology," *Science and Public Policy* 29, 6 (2002): 419-429

[48] *DOE and International Science and Technology. Synthesis Report on DOE's Contribution to Canadian Foreign Policy*, International Affairs Directorate, February 1983

[49] *Making the International Connection: How Canada's Approach to International Science and Technology can Help Small and Medium-sized Enterprises*

in the 1996 federal S&T strategy, *Science and Technology for the New Century*. It required departments to "develop specific plans to promote international collaboration for the benefit of Canadian firms." By way of response, EC's Science Policy Division drafted a report on the Department's action on the commitment. It referred to EC's business plan, the work on environmental technologies and the indirect benefits of its international collaboration in meteorology and hydrology for certain Canadian industries (e.g., shipping, aviation and agrifood).[50] No further policy work was done.

EC participated in 1998 in an interdepartmental effort, led by Foreign Affairs and International Trade, Industry Canada and the NRC, to develop an international S&T framework. But this made little progress. The Department was also involved in the initiative by the Prime Minister's Advisory Council on S&T on international S&T, which started in 1999. Part of the Council's report, like the 1994 National Advisory Board report, focused on small and medium-sized firms. However, the Council also called for the creation of an arms-length executive committee that would define Canada's international S&T priorities and coordinate international S&T activities. And it recommended a new fund be established to increase Canada's involvement in international S&T.[51] None of the Council's recommendations were acted on.

At the same time as the Council was delivering its report, Foreign Affairs and International Trade was rejuvenating its interest in S&T. It had closed its S&T Division at the time of Program Review and had considered eliminating its science counsellor positions abroad (in the end, the already relatively small number was reduced). In 2000, Foreign Affairs created an interdepartmental task group to provide direction to the science counsellors. The following year, the group's mandate was broadened to include other international S&T coordination issues. It was renamed the Interdepartmental Network on International S&T. EC has continued to participate in the Network, when it meets.

The May 2007 federal S&T strategy, *Mobilizing Science and Technology to Canada's Advantage*, also called for more to be done to encourage international collaboration. It committed the government to "assess Canada's S&T presence on the international scene and explore options to further improve Canada's ability to contribute to and benefit from international S&T developments, including through the Global

---

[50] *International Science and Technology Framework: Promotion of International Collaboration in Science and Technology for the Benefit of Private Firms at Environment Canada*, draft, April 1997. Science Policy Division files

[51] *Reaching Out: Canada, International Science and Technology, and the Knowledge-based Economy*, report of the Expert Panel on Canada's Role in International S&T, Advisory Council on S&T, 2000

Commerce Strategy." The interdepartmental ADMs Committee on S&T, which had been given responsibility for overseeing implementation of the strategy, tasked a working group (on which EC participated) to flesh out the commitment. The resulting report made a number of recommendations. It encouraged departments to align their international S&T with government priorities and to link it with activities in other sectors; to consider coordination across departments and the production of an annual report; and, to explore new funding models to integrate and leverage expertise in all sectors.[52] The report was provided as background information to an advisory body, the Science, Technology and Innovation Council, which was working at the time on Canada's involvement in international S&T.[53] Since the Council's advice to the government is considered private, no information is available at this time on its recommendations.

---

[52] *The Government of Canada's Engagement in International Science & Technology and the Path Forward*, draft report of the DGSI Working Group on International S&T, November 2008
[53] The Science, Technology and Innovation Council was created through the 2007 Strategy and reported to the Minister of Industry. It replaced two advisory bodies: the Advisory Council on S&T and the Council of S&T Advisors.

PART THREE

CAPACITY

4

Ensuring Capacity

As the inputs to S&T activity – money, buildings and equipment, technically trained people – S&T capacity appears to be a simple thing. Policy for ensuring capacity, however, has proven to be quite complicated. This chapter and the next two review that policy effort. This one focuses on funding, the others on capital expenditures and human resources. The assembly of data on S&T expenditures and reporting on them has been an ongoing activity of EC's policy for science effort. A summary of these expenditures can be found in Appendix 1. But that activity is not this chapter's subject. Rather, it looks at the general policy work to defend or increase EC's S&T capacity.

For most of its first 40 years, concern about funding EC's S&T activities was a constant issue. In the latter half of the 1970s and for most of the 1980s, it was at the root of deep dissatisfaction with the management of science in the Department. More recently, it has generated several public appeals by EC researchers.[1] However, S&T capacity and the funding necessary for it were rarely high up on the Department's policy agenda. This chapter examines the few occasions when it was – the early challenge of the government's Make-or-Buy policy, the preparations for the Green Plan, EC's strategy for dealing with Program Review, and the Department's attempts to assess its S&T capacity requirements and to argue for further resources in the years immediately following Program Review.

These episodes reveal the challenges faced in making a case for increasing EC's S&T. Although a fundamental question in policy for science, S&T capacity rarely achieved that status in department-wide deliberations within Environment Canada. The main reason for this was the subordinate position of S&T to EC's environmental goals, which were distributed among the highly autonomous parts of the Department. The most effective way of obtaining new resources for S&T turned out to be to make them part of a memorandum to cabinet for a new environmental program. This subordination and fragmentation of EC's S&T dampened the demand for departmental policy for capacity.

In addition to the situation within EC, there were other challenges. Determining the level of capacity that was required was inherently difficult. But the biggest challenge came from the federal government. Its struggles with deficits and its inclination to build up S&T capacity in the

---

[1] See, for example, *Beyond the Breaking Point?*, 2004

university and industry sectors, usually at the expense of federal departments, created a major barrier to the effectiveness of policy work in the area of S&T capacity. That work was necessary to define criteria for what S&T should be done by the Department, assess its S&T requirements and provide arguments for increasing capacity. But it was not sufficient to deliver the needed resources.

Make-or-Buy

Within a year of the creation of EC, the Science Policy Branch was assigned responsibility for guiding the Department's implementation of the government's new Make-or-Buy policy.[2] All departments were now required to contract out to industry their new mission-oriented R&D rather than carry it out themselves. The policy was aimed at raising the innovative capability of Canadian industry, thereby bringing additional economic benefits to the country.[3] Its formulation took place in the context of a widely held view that Canadian industry did too little R&D and that government performed too great a proportion of the total of Canadian R&D.

The policy was announced in 1972. The Science Policy Branch set up a separate unit, the Science Policy Analysis Division, to guide and coordinate the work of the Services in responding to the policy.[4] It took three years before the Division was closed and the Services were able to handle the policy on their own. Such central guidance was needed because of the variety of the Department's R&D and because there were some exceptions to the general requirement to contract out all new R&D. The government had indicated some principles and criteria to be used in justifying what R&D should be retained in government, but Treasury Board was slow to issue guidelines. The Division developed interim procedures, which were approved by the departmental management committee in December 1972.[5] Several criteria were listed in the 50-page document for keeping R&D in-house: where reasons of security prohibit industrial involvement; where no suitable industrial capability exists; where the R&D is essential to a regulatory function or to a set of national standards; and, where its conduct is essential to establishing and maintaining required in-house competence or to operating facilities which provide central testing and research services necessary to Canadian

---

[2] Memo, April 27 1976. LAC Acc. 1993-94/003, box 3
[3] See chapter 9 for a review of EC's policy work on industrial innovation.
[4] Memo, October 7 1976. LAC Acc. 1993-94/003, box 3
[5] Meeting, December 7 1972. LAC Acc. 1991-92/017, box 5

industry.[6] EC was reported to have played a key role in the application of the policy, with its guidelines being adopted by Treasury Board.[7]

Make-or-Buy had an impact on the Department's capacity. There was an increase in contracting out to industry. In 1971-72, about $2 million, 2% of EC's R&D expenditures, went to industry, mostly from the Lands, Forests and Wildlife Service.[8] Four years later this amount had risen to more than $8 million, over 7% of the total.[9] Although the policy affected a small share of EC's R&D, it touched many aspects of the Department's scientific effort: programs, the construction of S&T facilities, the financial management system, and the morale of its scientific staff.[10]

The policy achieved some success, but its results fell far short of its goal. Among the major barriers were the slow growth of funding for government R&D and a lack of alignment between government scientific missions and industrial capability.[11] A MOSST evaluation hinted that the government's applied scientific activities offered significant opportunities for industrial participation.[12] Indeed, some of EC's contracting out was in this area.[13] In an attempt to bolster the policy, the government issued a new version in 1977, called the Policy for Contracting-Out the Government's Requirements in Science and Technology. Not only did it expand the reach of Make-or-Buy to include applied scientific activities, it now moved beyond new S&T work to encompass ongoing activities.[14]

The new version created concern in EC. The Department supported the general principle of contracting out – it contracted out more than any other department – but it was apprehensive about the policy's impact on its ability to support its environmental mandate.[15] EC's doubts led to tension with MOSST. The latter complained to the Department's minister that it was being uncooperative.[16] As it had done at the beginning of the Make-or-Buy policy, EC created a unit in 1977 to

---

[6] *Interim Procedures to implement the Make-or-Buy Policy in Environment Canada*

[7] Memo, April 27 1976. LAC Acc. 1993-94/003, box 3. Treasury Board final guidelines were circulated in January 1973.

[8] Memo on R&D expenditures for 1971-72. LAC Acc. 1993-94/003, box 9

[9] *The Make-or-Buy Policy 1973-1975*, MOSST, 1975. LAC Acc. 1993-94/003, box 17

[10] Memo, January 11 1974. LAC Acc. 1993-94/003, box 1

[11] Peter Meyboom, "In-house vs contractual research: the federal Make-or-Buy policy," *Canadian Public Policy* 17, 4 (1974): 563-585. Many government scientists viewed Make-or-Buy as an indicator of the government's lack of commitment to its own scientific capacity.

[12] *Op. cit.,* footnote 7

[13] *Op. cit.,* footnote 9

[14] *Policy and Guidelines on Contracting-Out the Government's Requirements in Science and Technology,* Treasury Board, April 1977. LAC Acc. 1993-94/003, box 17

[15] Memo, June 17 1976. LAC Acc. 1993-94/003, box 3

[16] Meeting of departmental management committee, November 4 1976. LAC Acc. 1991-92/017, box 18

prepare the Department for Contracting-Out. The S&T Procurement Program was only needed for about a year. EC was having trouble in identifying additional activities it could contract out. The DM wrote to the Secretary of Treasury Board in January 1978 that most of EC's "S&T activities are inappropriate for contracting out."[17] He had the backing of his minister who was reported to be "quite concerned that the Contracting-Out Policy should not compromise the decisions of the Department on scientific work or the ability to obtain and retain good scientists."[18] The Department set up an ADM-level committee to identify eligible activities, but it too could not come up with much.

Although Make-or-Buy had increased the amount EC contracted out to industry, within six years departmental activities with commercial potential had plateaued. A study by the Economic Council of Canada in 1982 concluded that Make-or-Buy had failed in its aim of increasing the amount of government R&D contracted to the manufacturing sector.[19] It argued that the policy should have applied only to those departments where significant benefits could have been expected. EC, the study noted, was not among these. The Department's R&D had few spin-offs, and most of its contracts with industry were with the service sector. The Make-or-Buy policy was a spent force by the time of the Economic Council's study. The government was developing other programs for stimulating industrial innovation. However, contracting out or transferring government S&T remained a significant policy issue. It arose in many policy studies and recommendations and was usually a constant in discussions about S&T performed by government.[20]

The Green Plan

The Green Plan is usually remembered as a major environmental initiative under the Brian Mulroney Progressive Conservative government. But the Plan was also significant for its investments in S&T. Twenty-nine of its components had considerable S&T content, totalling about $1.2 billion – approximately one-third of the Green Plan's

[17] Letter, January 12 1978. LAC Acc. 1991-92/017, box 21
[18] Memo, January 13 1978. LAC Acc. 1993-94/003, box 5
[19] A. B. Supapol & D. G. McFetridge, *An Analysis of the Federal Make-or-Buy Policy*, discussion paper no. 217 (Ottawa: Economic Council of Canada, June 1982)
[20] For a brief overview see J. Kinder & F. Welsh, "Performing Strategic Science in the Public Interest: Updating the Policy Debate Regarding Government Science," *Scientia Canadensis* 35 (2012): 135-149

funding.[21] This large outlay for S&T provided a huge opportunity for EC to address its longstanding concern about the decline of its S&T capacity.

EC's S&T expenditures had grown in the decade of the 1980s (see Appendix 1). However that growth was mostly in the Department's non-research scientific activities (RSA). R&D expenditures had actually declined by 8%, 44% if calculated in constant dollars. The Department's science managers believed that there was a widening gap between ever-increasing demands on their science and its ability to deliver. A July 1989 workshop on EC's strategy for S&T went so far as to recommend that the Department seek an additional $500 million a year for its S&T and a 35% increase in S&T staff.[22] Planning for the Green Plan that year gave them an opportunity to deal with their S&T needs.

About 40% ($475 million) of the Green Plan's S&T was carried out by federal departments. How much of this came to EC has not been calculated. However, the data in Appendix 1 show an increase of about $73 million in EC's S&T expenditures in the early 1990s. This amount included a very large increase for R&D, which went up by $44 million in the same period. One of the main sources of this growth was the Green Plan initiative aimed at revitalizing research facilities and equipment and augmenting scientific staff.[23]

A memorandum to cabinet on "Revitalization of Environment Canada's Laboratories" was prepared in 1991. Its purpose was to "maintain the scientific capability necessary to deal with everyday environmental problems and to provide the scientific back-up required and expected by the other initiatives of the Green Plan."[24] Out of 41 proposals for funding, 17 were included in the MC. The total ask of the 17 was then cut in half to fit the $75 million available. The proposals covered the eight major research facilities in the Department, and included items such as air pollution measurement technology, global baseline surface radiation measurement, ecotoxicology and biochemical approaches to ecosystems management, and wildlife toxicology and chemical research.[25]

The $75 million boost to EC's S&T capacity was planned to be spread over 6 years from 1991-92 to 1996-97. However, the funds were soon affected by budget reductions. A year after it started, the program was cut

---

[21] *Overview of Science and Technology in the Green Plan*, April 1991. EC OSA, box 2

[22] *Vision, Values and Vehicles: A Strategic Approach for Environment Canada's S&T Programmes*, workshop report, Environment Canada Science Committee, July 1989. EC Roots, box 9

[23] *Canada's Green Plan*, page 150

[24] *Environmental Science Leadership: Revitalization of Environment Canada's Laboratories. Annex B: Supporting Documentation to Treasury Board Submission*, October 1991. EC OSA, box 2

[25] For a description of the facilities and their activities at that time, see the published report *Environmental Science and Technology: An Overview* (Environment Canada, 1993).

by about 8%. And the April 1993 budget announced further reductions to the Green Plan. However, all of these cuts were soon dwarfed by Program Review. It is difficult to trace how much money was eventually delivered to EC's labs under the Plan's revitalization commitment, but it probably amounted to no more than $20 million.[26]

The federal government's attempts to deal with deficits were not the only obstacle facing EC's policies regarding its S&T capacity. The extensive S&T expenditures in the Green Plan had also caught the attention of the Minister for Science, William Winegard. In January 1991, soon after the release of the Green Plan, he wrote to the Minister of the Environment:

> As you well know, the Green Plan constitutes one of this Government's most significant initiatives, not only for Sustainable Development but also for the achievement of this Government's related S&T and competitiveness objectives.
>
> I believe that a special approach is needed that lets us take full advantage of its potential on all of these fronts. I have already indicated key challenges which we should address. They include the demonstration of federal leadership and responsibility for S&T management, the effective use of existing programs and institutions, and the reconfirmation of this Government's policy on S&T.[27]

Winegard proposed a joint submission to cabinet on S&T and the Green Plan which would include a framework to ensure that management of the Plan's S&T would be consistent with federal S&T policy.

EC's Office of the Science Advisor worked with Industry, Science and Technology Canada and other science-based departments to develop a Framework for S&T in the Green Plan. It was to be used in the memoranda to cabinet on the various Green Plan components to assess their S&T programs. The Framework consisted of six principles, which were mostly based on the government's 1987 *Decision Framework for Science and Technology*. Both frameworks committed the government to: fostering partnerships among industries, universities and governments; leveraging federal resources; capitalizing on existing programs and expertise; delivering S&T through the private sector and universities where appropriate and feasible; and, promoting the development of highly qualified personnel. The Green Plan Framework also included a principle not found in the Decision Framework – ensuring an appropriate

---

[26] *Revitalization of Environment Canada Laboratories. Years one and two*, August 1993. EC OSA, box 2

[27] Letter dated January 22, 1991. EC OSA, box 2

distribution of resources from problem definition through to problem solution. The Framework was approved by Cabinet in February 1991.

The experience with the Framework illustrates the challenge posed by federal S&T policy at that time to departments trying to increase their S&T capacity. Not only did they need to deal with tight federal budgets and restrictions on the growth of the public service typical of efforts in the 1980s and early 1990s to reduce the deficit, but departments were also expected to support an economic development agenda. This meant using their S&T expenditures to form alliances with the private sector and to increase R&D in universities and in the private sector. Even Environment Canada, which was not primarily an economic department, faced these pressures.

Program Review

The promise of increased S&T capacity through the Green Plan was short-lived. The new Liberal government announced its intention to undertake a program review in the February 1994 Budget. This would have a substantial impact on EC's S&T over the next five years, more than reversing the gains under the Green Plan.[28]

EC's strategy for dealing with Program Review was to maintain core Departmental expertise at the expense of extramural funding. Consequently, funding for university research went from $16.8 million in 1994-95 to $2.4 million in 1997-98. EC management also decided to increase its partnerships with others. A good example of this is the effort put into the MOU between the Four Natural Resource Departments on S&T for Sustainable Development (see chapter 7). In addition, EC would work to harmonize its efforts with the provinces, build the capacity of others by sharing information and encouraging the development of new technologies, use automation and technology to reduce costs especially for weather forecasting, and provide specialized services on a commercial or cost recovery basis.[29]

In responding to Program Review, EC's senior officials had decided to place relatively more emphasis on science and on policy, and less on service delivery.[30] However, there was little room to manoeuvre in a

[28] For an overview of EC's handling of Program Review, see Donald J. Savoie, "Towards a Different Shade of Green: Program Review and Environment Canada," in *Managing Strategic Change: Learning from Program Review*, eds. Peter Aucoin and Donald J. Savoie (Canadian Centre for Management Development, 1998), 71-97.
[29] *Environment Canada. Program Review. Implications for Science*, Environment Canada Science Committee, February 23 1995. EC S&T Management Committee, box 1
[30] *Deputy Minister's Message to all Employees*, January 23 1995. EC OSA, box 6

budget dominated by S&T. EC's expenditures on S&T were reduced by 36% from 1993-94 to 1998-99 (Appendix 1). The number of person-years devoted to S&T declined 39%, some 1,818 person-years (Appendix 2). R&D, however, fared much better. Its expenditures declined by 14% in the same period, person-years devoted to it by 4%. In 1993-94, R&D had been 20% of S&T expenditures. By 1998-99 it had risen to 27%.

Program Review not only had an impact on the level of resources for S&T, it also took up the time of senior managers. Their attention was directed to administering cuts rather than building programs. This situation began to change towards the end of 1997. Less preoccupied with downsizing and encouraged by the much-improved federal budgetary situation, the Department started policy work on whether it had the resources needed to deal with future challenges.

S&T Capacity in 2002

When EC began to turn its attention to whether it had sufficient S&T capacity, the focal point of this work was the newly created R&D Advisory Board. At its second meeting, in October 1997, the Board was presented with six key issues facing EC's S&T. Among these was S&T capacity. The Board decided to set up a working group to look at whether EC would have the S&T it would need in five years' time – in 2002.

The working group, as was the Board's practice at that time, consisted of both Board members and EC employees. Suzanne Fortier, the Vice-Principal for Research at Queen's University, and David Neave, the Executive Director of Wildlife Habitat Canada, represented the Board. They were joined by about ten EC employees, including staff from Science Policy. Over the following year, the working group held a series of meetings to review and discuss material on the Department's mandate, S&T expenditures, research publications and demographics. It also sponsored a one-day brainstorming session, involving several departmental S&T managers, on the challenges in ensuring adequate S&T capacity.[31] In a report back to the Board in October 1998, the group summed up its views to that point. It believed that EC's S&T workforce posed no immediate problems, that the statistics on expenditures and publications it had reviewed were useful but that more management information was required, and – most significantly – that determining

---

[31] "Coming to grips with environmental S&T capacity in Canada for the year 2002 – what are the challenges?" Discussion paper, September 24 1998. EC S&T Advisory Board, box 5

EC's priorities was key to capacity assessments. The report met with general agreement.

The working group then organized a two-day workshop in February 1999. It focused on understanding the current and future business of the Department (the DM participated in this session), identifying gaps in S&T capacity for 2002, and developing a set of recommendations to address immediate needs and to establish a long-term framework for S&T capacity assessment. Based on the workshop and its previous work, the working group presented its findings to the Board in March. These were adopted by the Board and transmitted to the DM.

In the report to the DM, the Board noted that there appeared to be gaps in EC's ability to carry out its tasks in certain areas, while in others resources were spread very thin.[32] The Board concluded that this situation needed to be better documented and substantiated. Its main recommendation was for an external review of EC's S&T capacity to address emerging and future environmental issues. The Board also made several short-term and long-term recommendations.[33] It advised the Department to immediately review and take corrective action on vulnerable areas of scientific activity, to inventory capital assets and develop a rationalized capital plan for its S&T, and to ensure that past successes and current information on its S&T were communicated to the public. In the longer term, EC was urged to draw up a clear statement of its core mission, to link its business lines to that, and to develop science agendas for those lines. The Department was also asked to take into consideration, in those agendas, environmental S&T capacity outside EC; to focus greater effort on strengthening external S&T capacity; to make more visible its partnerships; and, to develop a longer-term planning process.

In his response to the Board, the DM thought that the recommendation for an external review was premature.[34] He believed that the other recommendations should be worked on first. He listed various EC initiatives related to the recommendations. For example, negotiations were in progress for new space for the National Wildlife Research Centre, a long-term capital plan had been developed, an S&T partnering framework approved, an S&T home page established on EC's

---

[32] This view was shared by other knowledgeable observers. See, for example, G. Bruce Doern, "Patient Science versus Science on Demand: The Stretching of Green Science at Environment Canada," in *Risky Business: Canada's Changing Science-Based Policy and Regulatory Regime*, eds. G. Bruce Doern and Ted Reed (University of Toronto Press, 2000), 286-306.
[33] *Achievements to Date, Recommendations for Future Action*, S&T Advisory Board Report No. 1, March 2000
[34] The Meteorological Service did conduct an external peer review of its R&D in 2001: *2001 PEER REVIEW Research and Development Program*.

website, business plans identifying results and priorities developed for the new business lines, and research agendas were being prepared for the business lines.[35]   That EC was already taking action on the Board's recommendations should not be surprising. Because the Board operated through working groups that included many Departmental staff, the Board's thinking reflected the ongoing and planned activities within EC to reinvigorate its S&T capacity. In turn, the Department did not have to wait for formal recommendations before moving forward.

The findings of the Board highlighted the difficulties in making the case for increased S&T capacity. There was no simple way to calculate how much or what capacity would be needed. Many inter-related factors were involved – such as short- and long-term goals, external capacity, emerging issues, changing currents in S&T, existing human resources, and facility and equipment needs. It was also difficult to roll up requirements for specific issue areas into a general case for S&T capacity. Even so, the efforts of the Board and its working group helped the Department to think through the issue of S&T capacity. The DM acknowledged that contribution when he wrote to the Board:

> I am pleased to tell you that, in numerous instances, the Department is already addressing many of the Board's recommendations and we are initiating action on several of the others. The recommendations reinforced the need for our attention to some important issues, helped focus some activities already underway, and have made an important contribution to the development of an overall approach for assessing S&T capacity within Environment Canada.[36]

Working with Others to Make the Case for S&T

The R&D Advisory Board's work marked an end to general discussions about S&T capacity in EC. The Department turned its attention to obtaining funding, which now seemed increasingly possible due to federal surpluses. Working within the conditions set out by Treasury Board's Program Integrity exercise, which started up in 1998, the Department focused first on infrastructure needs and then on those of the meteorological service and of the Canadian Environmental Protection Act. It also took advantage of specific opportunities for new funding that arose through, for example, the Toxic Substances Research Initiative, the Canadian Biotechnology Strategy Fund for Collaborative Projects, and

---

[35] *Op. cit.,* footnote 33
[36] Ibid., letter, September 29 1999

the Climate Change Action Fund. This led to a reversal of the decline, caused by Program Review, in both the Department's overall budget and its S&T expenditures (see Appendix 1).

Obtaining increased S&T resources was not an easy task. A case in point is provided by the result of the October 1999 Speech from the Throne. It had included a commitment to strengthen the government's environmental S&T capacity. When the Budget was released the following April, there was a $500 million boost for environmental S&T. However, almost all of this went to others, in particular to non-governmental groups – e.g., $100 million for the Sustainable Development Technology Fund and $60 million for the Canadian Foundation for Climate and Atmospheric Science. Only a very small amount went to Environment Canada, mostly for S&T work on species at risk. The Department's S&T needs were not a high political priority compared to those of universities and the private sector.

EC's policy efforts in S&T capacity shifted its focus to gaining allies outside of the Department. In one stream of activity, it began to explore the broader, Canadian environmental research enterprise. The interest here was in using new models to organize and mobilize this capacity, and in determining EC's role in this.[37] (This subject is explored further in chapter 8.)

The second approach was to work with other science-based departments. Many of these were also preoccupied with the issue of S&T capacity.[38] Several of them, including EC, decided to work together under the auspices of the ad hoc committee of ADMs on science in government.[39] The committee organized an interdepartmental workshop in February 1999 to share information and best practices, to ensure compatibility of information, to identify and close gaps in that information, and to develop an action plan for a government-wide policy paper on S&T capacity. During the remainder of 1999 and early part of 2000, the committee was largely focused on trying to make a common case for investment in federal S&T. Presentations were drafted and made to DMs and to Cabinet. But these suffered from trying to please all the parties involved. They ended up being mostly about increasing awareness of the issue and did not offer a way forward.

---

[37] See, for example, *Strengthening Environmental Research in Canada: A Discussion Paper*, The Impact Group. Science Policy Branch working paper #5, December 1999

[38] *Federal Science Capacity: Compilation of Departmental Studies*, report by Hickling Arthurs Low Corporation, February 1999. EC ADMs' Committee on Science in Government, box 1

[39] This was an informal, small committee of ADMs from the major science-performing departments. It had started meeting in August 1997 in response to the months of negative press that summer about federal science (see chapter 10). It was chaired by Marc Denis Everell, then the ADM for Earth Sciences and Chief Scientist at NRCan.

The ad hoc ADMs committee also took some other steps to raise the profile of issues faced by federally performed S&T. It pressed Industry Canada to implement the commitment in the 1996 federal S&T strategy to create an external advisory body focused on S&T in the federal government. Another commitment, the Advisory Council on S&T, which was devoted to S&T issues outside the federal government, had already been in existence since 1996. The result was the formation of the Council of Science and Technology Advisors. Composed of representatives from the external advisory bodies of science-based departments, it had its first meeting in June 1998. The Council soon turned its attention to the issue of S&T capacity. Its 1999 report, *Building Excellence in Science and Technology (BEST): The Federal Roles in Performing Science and Technology*, looked at the roles of the government in performing S&T and at its capacity to deliver on them.

In the report, the Council stated that it believed that there was a critical role for the federal government to play in performing S&T.[40] The Council also believed that the S&T capacity problem was real. However, like EC's R&D Advisory Board, it stressed that the challenge was not in restoring capacity to historic levels, but in identifying "what capacity is needed to allow the government to meet current needs and enhance its ability to meet future challenges." The Council did not like the approach many departments tended to take of doing the same with less.[41]

The efforts of the ad hoc ADMs committee on science in government to take a collective approach to the issue of federal S&T capacity did not meet with much success. While some progress was made in raising awareness, federally performed S&T did not share in the large increases in government expenditures for S&T. Changing political perceptions of departmental S&T would take more than a few overview presentations. In addition, despite many calls for a strategic approach to federal S&T investments, a government-wide approach proved to be too removed from specific interests. Whatever funds federal science-based departments successfully obtained mostly came through individual departmental priorities.

There was not much reason for the ADMs to keep meeting as a distinct group. By the summer of 2000, the committee had ceased to meet. However, other capacity initiatives dealing with infrastructure and human resources still continued to be pursued (these are dealt with in the following two chapters). And a new ad hoc interdepartmental ADMs

---

[40] The report laid out four key roles for federally-performed S&T. These were later adopted by the federal government in its *In the Service of Canadians: A Framework for Federal Science and Technology*, 2005

[41] *Building Excellence in Science and Technology (BEST)*, p. 40

committee would soon arise with another approach to making the case for federal S&T, one designed to be innovative and appealing to central agencies and politicians. Led by EC, this one would seek increased departmental S&T capacity by creating synergies through enhanced collaboration among science-based departments (see chapter 7).

5

Providing Infrastructure

Infrastructure is one of the critical elements of EC's S&T capacity. It encompasses facilities and land holdings, vehicles, scientific equipment and information technologies. Because of the multiple functions played by EC's S&T, its infrastructure is varied, extensive, sometimes large-scale, and often not widely available elsewhere in Canada. The nature of the government's S&T infrastructure is one of the main features distinguishing S&T in government from S&T in other sectors. It is an important factor in the government's ability to attract S&T talent to its service.

Similar to other aspects of capacity, the main challenge facing EC's policy for S&T infrastructure was financing. It was easier, especially when resources were tight, to reduce expenditures on maintenance and new equipment than to dismiss staff or cut back on operations. The short-term tactic had consequences. Over time, it was increasingly difficult and costly to maintain S&T infrastructure at acceptable standards and to keep up with shifts in environmental issues and the emergence of new technologies.

Obtaining funds for S&T infrastructure faced many hurdles. The government had to have funds and the issue had to be a priority. EC needed to take into consideration future requirements and possible synergies, and not simply the replacement of existing facilities and equipment. This required the involvement of both capital asset and S&T managers. In addition, the issue had to be a major concern for the Department. However, for most of EC's history this was not the case. Despite the issue's importance to operations, the Department had no overall plan or management system for infrastructure. It was left up to the Services to manage and rarely was a topic for the attention of the departmental management committee.

EC's policy work on S&T infrastructure occasionally extended beyond the Department, particularly after Program Review in the mid-1990s. Interdepartmental coordination held out the promise of greater benefits in the use of those assets. However, the collective approach did not prove to be successful. Working together had additional transactional costs. And central agency leadership and policies in support of interdepartmental coordination were absent. Policy for S&T infrastructure would remain squarely in the domain of individual departments.

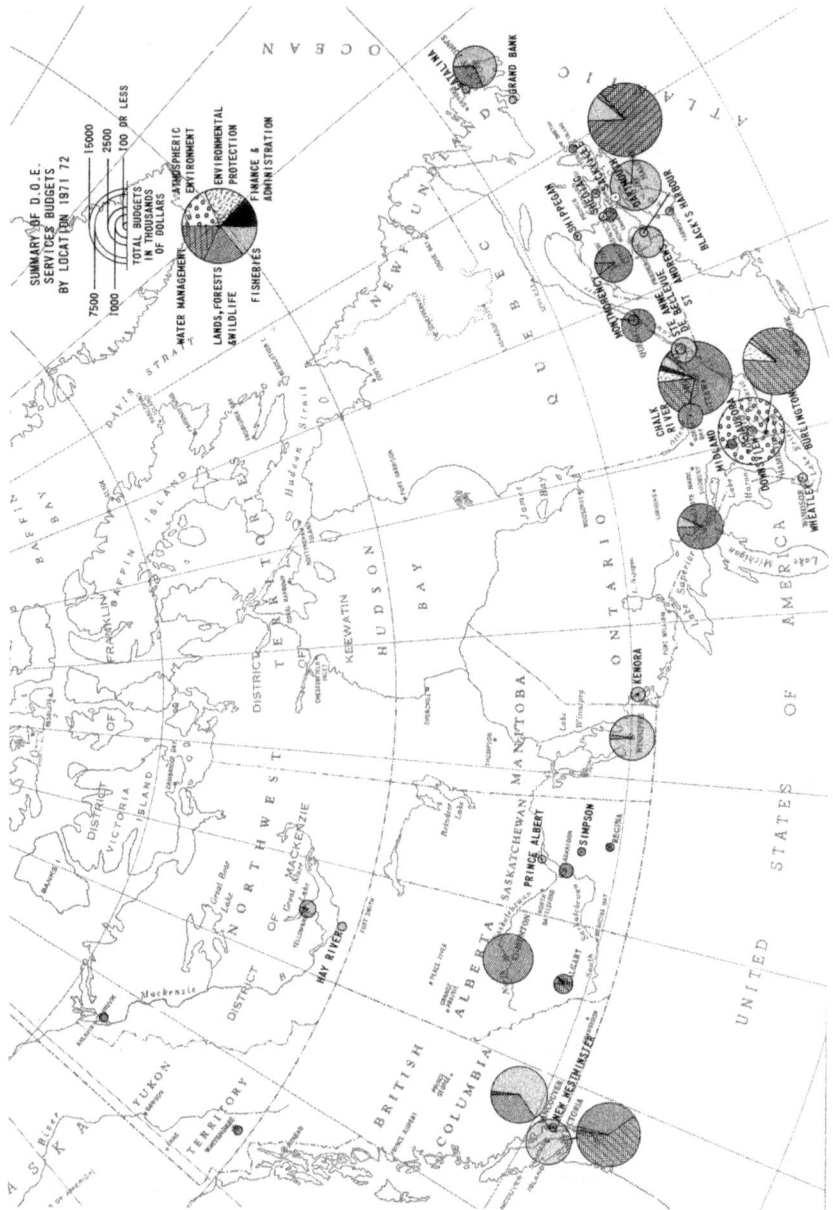

**Figure 5.1 EC Service intramural S&T budgets
by location, 1971-72[1]**

---

[1] *Compendium of Scientific Establishments: Summary*, December 31 1971

Task Force on Research Facilities

EC was engaged in considerable discussion in the fall of 1971 around plans to build new research institutes on the West Coast, in the Arctic and in Hull. This led to several studies and task forces. One was a paper on the major science policy issues facing the Department.[2] Another was the *Compendium of Scientific Establishments*, which presented an overview of EC's scientific facilities and their activities.[3] A third was the establishment of a departmental task force on new facility requirements. A request from Treasury Board for capital projections had helped to spark its creation. Peter Meyboom, the Director of Science Policy, chaired the task force. Its mandate was to project capital requirements up to 1977-78 for research facilities and ships as well as for operational labs. It was also to draw up criteria to rank proposals for these expenditures, to suggest priorities and to recommend a five-year plan for capital expenditures.[4]

The task force reported in March 1972.[5] It ranked 16 proposals for research and operational facilities, amounting to a total of just under $130 million in construction costs. The departmental management committee generally agreed with the report. However, it wanted to review the ranking in the light of the department's program priorities.[6] This proved difficult to do.[7] In the end, each Service was left on its own to plan and find financing for its capital needs. As had happened with several other department-wide initiatives in the early history of EC, decision-making on the need for S&T facilities was ceded to the individual Services.

Operational S&T

Monitoring

Monitoring has always been an important part of EC's scientific activities. In 1971, the Department operated networks for collecting "meteorological, hydrometrical, hydrochemical, hydrographic,

---

2 See chapter 2.

3 The *Compendium* consisted of two volumes, a *Summary* and an *Appendix* that provided details by Service. Another version, *Compendium of Scientific Establishments in Environment Canada*, was produced in August 1973 to reflect the reorganization of the Department earlier that year. LAC Acc. 1993-94/003, box 26

4 Memo, December 21 1971. LAC Acc. 1983-84/141, box 12

5 *Report by the Cross-Mission Task Force on Research Facilities*, March 24 1972

6 Meeting, April 6 1972. LAC Acc. 1991-92/017, box 3

7 Meeting, April 19 1972. LAC Acc. 1991-92/017, box 3

sedimentological, biological and air quality data."[8] This effort generated several issues: what would it cost, to what extent would it make use of new technologies and automation, was sufficient measurement taking place, were the right things being measured, should different approaches be taken, and how should the data be managed and made available to the public. Despite the many issues, monitoring rarely became a major policy concern for the Department.

In June 1972, Peter Meyboom presented a discussion paper on environmental monitoring to the departmental management committee.[9] It listed the main federal-provincial and organizational issues involved.[10] However, the committee took no action on the paper. Two years later, the Science Policy Branch was asked to prepare an inventory of the Department's monitoring activities. *Environmental Monitoring: A Compendium of Data Gathering Activities of Environment Canada* was organized by Service. The Atmospheric Environment Service, for example, was reported to be spending $26.2 million on the collection of atmospheric and ice data and $18.4 million to process, analyze and provide forecasts – almost the entire budget of the Service. At the same time, the Department received an unsolicited proposal to design an integrated system for monitoring the physical and biological environment. It was prepared by a consortium of private sector firms led by Philips Electronic Industries. The Department agreed to support the study, which resulted in a four-volume report, *Environmental Monitoring in Canada*.[11] It urged the various levels of government in Canada to create a fully integrated and comprehensive environmental management system. It recommended a federal-provincial conference to work out a national and coordinated approach to monitoring, as well as a Canadian Environmental Monitoring Commission to set objectives and priorities for the system. EC set up a group to review the report. It found it "difficult to conceive of an effective fully integrated monitoring system" given the needs of different governments, and did not believe such a system to be "desirable or necessary".[12] It also did not favour the idea of a Commission. The group recommended that the report not be published.[13]

The reaction of the review group revealed the challenges within EC in developing a departmental policy on monitoring. The heterogeneous

---

[8] EC, *Science in a Changing Environment – proposals for a departmental science policy*, January 1972

[9] *National Networks, Environmental Monitoring, Surveys and Surveillance*. I have not located a copy of the paper. Reference in LAC Acc. 1991-92/017, box 4

[10] *Environmental Monitoring: A Compendium of Data Gathering Activities of Environment Canada*, 1975

[11] LAC Acc. 1993-94/003, box 22

[12] Memo to DM, July 28 1978. LAC Acc. 1993-94/003, box 22

[13] Memo, August 3 1978. LAC Acc. 1993-94/003, box 5

nature of monitoring coupled with very autonomous Services made the development of an overall plan for environmental monitoring very difficult. Yet the scale of monitoring activities and the need to deal with new environmental issues meant that concern about monitoring remained. Although usually an undercurrent, it did occasionally rise to the attention of senior management. For example, the Department produced a report on integrated monitoring in 1990, it maintained a coordinating office for an ecological monitoring and assessment network from 1994 to 2010, and development of an integrated environmental monitoring and prediction capability was the first strategic direction in its 2007 *Science Plan.*[14] While formulating policy for monitoring proved elusive, developing plans for specific monitoring systems was an ongoing activity, as illustrated recently in the preparation of an environmental monitoring plan for the oil sands.[15]

Analytical Laboratories

The Department's operational, or analytical, laboratories are one area of EC's S&T infrastructure that has long attracted management attention. They provided an essential service by testing environmental samples and undertaking other related activities such as analytical research and methods development, quality assurance and quality control, expert testimony, field support, and contract management. Many studies were conducted over the years to improve their coordination and optimize the services they supplied. It was budgetary constraint that usually drove closer examination of the labs.

Meyboom's 1972 paper on research facilities had recommended that the relationships among EC's water quality labs be reviewed. They were located in three Services: Environmental Management, Fisheries, and Environmental Protection. That April, the departmental management committee established a working group to look at how the labs might be more closely connected. Based on the subsequent report, the management committee decided in November that the chairs of EC's Regional Boards would be responsible for "improving laboratory co-ordination, laboratory technical and workload effectiveness and other functions" for the water analytical labs in their regions.[16] It also

---

[14] *A DOE Management Framework, Strategy and Action Plan for Environmental Monitoring,* 1990
[15] *An Integrated Oil Sands Environment Monitoring Plan,* 2011
[16] *Report of the Working Committee on Interservice Coordination of Analytical Laboratories in the Prairie and Northern Region,* April 1973

appointed a departmental coordinator to help the chairs develop a departmental policy.[17]

This approach to coordination of the labs had some initial success, but its management system soon lapsed. In May 1977 the DM requested that each region set up coordinating committees for lab and field activities.[18] The A-base review, which started that same year, also looked into duplication in the analytical labs and recommended a regionally integrated system. However, the growing demand for analytical services kept the issue alive. In early 1981, the DM charged the ADM of the Environmental Conservation Service with a review of departmental lab policy.[19] He wanted advice on the purchase of costly lab equipment, on the resource implications of new and changing priorities, and on co-location of the labs. The ADM reported back later that year.[20] He proposed a set of guidelines for the functions of the analytical labs and recommended some consolidation in both his and the Environmental Protection Service.

Following the report, the labs of the Inland Waters Directorate (part of the Environmental Conservation Service) were reorganized in 1982. This involved centralization of complex and expensive analyses at the National Water Quality Laboratory in Burlington, while maintaining regional labs for more routine work.[21] Environmental Protection's labs in the Ottawa area were consolidated at River Road. That Service also set up a Laboratory Managers' Committee for its labs. When the Environmental Protection and Conservation Services were merged a few years later, in 1986, the Committee expanded to include all the labs in the new Service.

The Conservation and Protection Service undertook another review of its analytical operations in 1988, in order to reduce the person years invested in that activity. At the time, the Service had 10 operational labs reporting to regional offices and 4 others to headquarters, involving 150 person years, between $3 million and $4 million in operations and maintenance funding, and $2 million in capital funding.[22] The Laboratory Managers' Committee was tasked with implementing the results of the review, which included clarifying the roles of the labs, finding a balance between them and the research laboratories, and establishing a

---

[17] Meeting, November 1 1972. LAC Acc. 1991-92/017, box 5

[18] *An Integrated Approach to the Provision of Laboratory Services for the Water Quality Programs in the Department of the Environment*, December 20 1978. EC Roots, box 9

[19] Meeting, February 25 1981. LAC Acc. 1991-92/017, box 36

[20] Meeting, September 16 1981. LAC Acc. 1991-92/017, box 38

[21] Meeting, January 27 1982. LAC Acc. 1991-92/017, box 38. See also *Improving the Science Base: A Report on Laboratory Analytical Services*, February 1989.

[22] *C&P National Laboratory Review: Action Plan Studies. Final Report*, December 1991

relationship with the recently formed Canadian Association for Environmental Analytical Laboratories.[23]

The Committee was also charged with reviewing and advising on plans for equipment replacement and for new acquisitions for items costing more than $100,000.[24] The role was strengthened when the ADM of Conservation and Protection set aside about $1 million per year in capital funds for the laboratories. The Committee helped determine where the funds would be spent.[25] It proved its usefulness by coordinating lab purchases and getting the most out of available funding. However, the Committee's work in this area was displaced a few years later by the Green Plan's laboratory revitalization initiative. The latter provided $75 million over 6 years to 17 projects aimed at updating both R&D and operational laboratory facilities and equipment.[26] The Committee did not play any role in reviewing the proposed projects, probably due to the timeframes involved, the inclusion of research laboratories, and because its membership was drawn from a lower level of management.

The Laboratory Managers' Committee continued to pursue its other functions. During the 1993 reorganization of the Department, it expanded to include similar labs in the rest of the Department – in the Atmospheric Environment Service and the National Hydrology Research Institute. At the time of Program Review, the notion of a group that would coordinate and provide advice on laboratory science for program delivery once again came forward. As part of an ongoing search to cut costs and in response to Program Review II, the Departmental management committee decided, in November 1995, to look for efficiencies in its laboratory services and set a target of $2 million in savings.[27] At that time, EC was spending about $19.2 million in this area, with 60% of the total for testing.[28] A laboratory review group was established (Figure 5.2 lists the laboratories it examined). It first looked at the supply and demand for laboratory services and then at options for improved efficiencies.

---

[23] The Association was established in 1989. In 2008 its name was changed to the Canadian Association for Laboratory Accreditation.

[24] *Laboratory Rationalization. Final Report*, prepared by Alison Kerry, April 1997. Science Policy Division files

[25] Ibid.

[26] For further information, see chapter 4.

[27] Program Review II was launched in 1996 as a second phase of Program Review.

[28] Op. cit., footnote 24. In 1995-96, EC required approximately 1. 9 million laboratory analytical results. The majority (77%) of these were provided by EC's laboratories.

ENVIRONMENT CANADA'S LABORATORIES

ENVIRONMENT CANADA'S INSTITUTES & LABORATORIES	REGION/ SERVICE	LEASED/ OWNED	LOCATION
National Research Institutes' Laboratories:			
National Water Research Institute (NWRI) • Aquatic Ecosystem Protection • Aquatic Ecosystem Restoration • Aquatic Ecosystem Conservation	ECS	Owned	Burlington, Ontario
National Hydrology Research Institute (NHRI) • Central Analytical Laboratory • Stable Isotope Facility • Organic Analysis Laboratories	ECS	Owned	Saskatoon, Saskatchewan
National Wildlife Research Centre (NWRC) • Wildlife Toxicology	ECS	Owned	Hull, Quebec
Environmental Technology Centre (ETC) • Analysis and Methods • Mobile Sources Emissions • Emergencies Engineering • Emergencies Science • Microwave-Assisted Process • Air Toxics Measurement	EPS	Owned	Gloucester, Ontario
Climate & Atmospheric Research Directorate (CARD) • 1 Lab Facility with 11 program related labs (toxics and air constituents) [labs at York University collocating here]	AES	Leased	Downsview, Ontario
Centre Saint-Laurent (CSL) • Aquatic Contaminants • Ecotoxicology & Environmental Chemistry	Quebec	Leased	Montreal, Quebec
Operational Laboratories:			
National Laboratory for Environmental Testing (NLET) in NWRI	ECS	Owned	Burlington, Ontario
Newfoundland District Laboratory [1 person sharing space in DFO Food Inspections Laboratory (under MOU)]	Atlantic	DFO Owned	St. John's, Newfoundland
Environmental Quality Laboratory - [BIO Toxicology lab, Dartmouth collocating here]	Atlantic	Leased	Moncton, New Brunswick
Water Quality (Nutrient) Laboratory (in NHRC)	PNR	Owned	Saskatoon, Saskatchewan
Sediment Laboratory (EP) - [to be collocated with another lab in PNR]	PNR	Leased	Regina, Saskatchewan
Environmental Protection Analytical Laboratory - [Edmonton Ecotoxicology/Bioassay lab collocating here]	PNR	Leased	Edmonton, Alberta (Northern Forestry Centre)
Pacific Environmental Science Centre	P&YR	Owned	North Vancouver, British Columbia

Figure 5.2 **EC's research and operational laboratories in 1997**[29]

The group reported back that the targeted savings could be found in laboratory services. It also recommended creation of a Laboratory Coordinating Committee that would "provide ongoing departmental coordination and management advice regarding the laboratory science activities necessary for program delivery, in order to make the most

---

[29] Op.cit. (footnote 24)

effective and efficient use of these resources nationally."[30] It would build on the activities of the Laboratory Managers' Committee but be "more strategic, inclusive and representative." The departmental management committee accepted the recommendation in May 1997. The Laboratory Coordinating Committee reported to the S&T Executive Committee (see chapter 1) and was chaired by a DG from the Environmental Conservation Service (whose ADM was the EC lead for science policy).

Capital planning was intended to be one of the main areas of activity for the new Committee.[31] But it is not clear whether it spent much time on this. The Committee's reports to the S&T Executive Committee focused on the accreditation of EC's laboratories and on the development of a departmental policy on laboratory data quality assurance.[32] There was some concern that the Committee's membership was not senior enough to be able to achieve its mandate. The suggestion was made that the S&T Management Committee should be responsible for many of its activities. In the 2005 reorganization of EC, the Laboratory Coordinating Committee disappeared. Most of the operational laboratories were consolidated into the Emergencies, Operational Analytical Laboratories and Research Support Division of the Water S&T Directorate.[33]

Treasury Board Secretariat Initiatives

As noted in the previous chapter, EC began to take stock of its capacity following the period of Program Review. One of the Department's main concerns was the effect of years of underinvestment in its capital assets. The issue was common to many other federal departments. So it is not surprising that one of the main components of TBS's Program Integrity initiative was to look at the rust-out of departments' facilities and equipment.

In early 1999, EC was asked for information on its capital needs. The Department reported back to TBS that its asset base was valued at $866 million.[34] Annual capital spending was only about $25 million at that

---

[30] Op. cit., footnote 24

[31] The others were rational distribution of laboratory work, sharing of best laboratory management practices, and ongoing review of the impact of new departmental priorities on laboratories.

[32] The policy was approved by the departmental management committee in 2003.

[33] The few others were located in the Atmospheric S&T and the Wildlife and Landscape Science directorates in the S&T Branch.

[34] *Capital Rust-out and Long Term Capital Plan,* June 17 1999. EC S&T Management Committee Meetings, box 4

time.[35] This was judged to be very inadequate to maintain the assets, let alone to deal with the need for new technologies and shifts in research priorities. The Department asked for $246 million to bring its asset base to an acceptable standard and another $52 million in ongoing funding to maintain that base. S&T infrastructure made up the majority of this request. At that time, EC operated 15 research institutes and laboratories, was responsible for 49 national wildlife areas, and had over 4600 air, climate and water monitoring stations across the country.[36] The Department ended up receiving less than 10% of its request, with the new resources being devoted to the move of the National Wildlife Research Centre to a new building on the Carleton University campus, and to health and safety requirements at certain meteorological stations and in the hydrometric network.

The lesson EC learned from this experience, as well as from other attempts to make the case for increased S&T capacity, was that the numbers in themselves were insufficient. The Department needed a strong storyline with a unified set of requests. It tried to deliver this in 2000, in its first long-term capital plan in a decade.[37] In addition, EC had to go beyond appealing for the replacement of existing infrastructure and to show how the infrastructure was needed for current and, more challengingly, for future departmental priorities.[38]

The standard procedure for departments seeking capital funding was to deal individually with TBS. The paucity of such funding led, as it had in other areas of capacity, to explorations of what might be done in collaboration with other departments. The Real Property Advisory Committee of TBS established a working group in March 2002 to discuss opportunities for partnerships. It was composed of corporate ADMs and DGs from the departments that were party to the Memorandum of Understanding on Science for Sustainable Development.[39] That summer, members of the working group made presentations to the science ADMs' committee that was developing FINE.[40] As a result, that committee identified several opportunities for a more integrated approach to developing and using federal facilities; for example, a GeoScience Centre in Quebec City, a Canada Water Centre built around the National Water Research Institute in Burlington, and a science and marine park at the

---

[35] It averaged $35 million per year for the decade 2000-2009.

[36] *Environment Canada's Long Term Capital Plan for the period 2000-2001 to 2004-2005*, August 2000

[37] Ibid.

[38] Such was the advice from EC's R&D Advisory Board and from the Council of S&T Advisors 1999 report *Building Excellence in Science and Technology (BEST)*.

[39] See chapter 7.

[40] Ibid.

Bedford Institute for Oceanography in Dartmouth. However, with the demise of the FINE initiative, none of these ideas were further developed.

Interdepartmental discussions about a more collaborative approach to S&T infrastructure got a boost a year later when the government announced the creation of a Cabinet committee on expenditure review. Capital assets were identified as one of the areas for review. They accounted for $3.7 billion of government expenditures in 2002-03.[41] The objectives there were to "identify potential savings or efficiencies around the management and use of the government's capital assets", to determine the barriers to that management and suggest solutions, and to recommend ways of improving "oversight and accountability for capital assets."[42] The review of capital assets was focused on four asset classes that involved many departments, one of which was S&T infrastructure.[43]

To support this S&T focus, a Science Infrastructure Review Working Group was set up by TBS in the spring of 2004. Although the group had originated in corporate management circles, it quickly sought out the help and advice of science ADMs. Its membership then included scientists and science policy personnel as well as capital asset staff. These were drawn from 12 science-performing departments and agencies, with its core coming from the five departments who had formed the earlier working group under the Real Property Advisory Committee. The working group kept in close touch with and sought leadership from the science ADMs Integration Board.

While the working group's main function was to serve the capital assets review through gathering and analyzing data on science infrastructure, it also incorporated the interests of its predecessor in increasing interdepartmental collaboration. Its mandate was:

> To improve the efficiency and effectiveness of the federal S&T capital asset portfolio through greater collaboration between the participating organizations, and through the implementation of partnership initiatives, both within the federal S&T community and with non-federal government S&T organizations.[44]

---

[41] *Capital Assets. Expenditure Review. Terms of Reference*, draft presentation, March 30 2004. Science Policy Division files

[42] Ibid.

[43] The others were office and warehouse space, vehicles, and surplus land.

[44] *Science Infrastructure Review Working Group. Terms of Reference*, April 14 2004. Science Policy Division files

The members of the group believed in the importance of S&T infrastructure in enabling the future directions of government science. And they held that significant improvements in the efficiency and effectiveness of that effort were possible through synergies or integration of the infrastructure.

As information on S&T infrastructure was not regularly collected in departments much less across the federal government, the group's first task was to gather data. It organized an inventory of federal S&T facilities and major pieces of equipment (those costing over $250,000). The inventory had to be done quickly over the summer because of the fall deadline for the capital assets review. The result was the first common database of federal S&T infrastructure. The group also assessed the state of those assets. It judged that two thirds of federal S&T buildings were in less than good condition (80% in EC) and that one third of the equipment it had inventoried was more than 10 years old (40% for EC).[45]

The working group recommended establishing an ongoing governance system for S&T infrastructure through a committee of S&T and corporate ADMs. It would share and integrate departmental priorities, projects and capital asset investment plans. The group also urged the continued development of the database it had created. And it recommended the creation of a framework for S&T infrastructure partnerships, as well as an assessment of the legal and administrative policy barriers to effective collaboration.[46]

The working group had hoped to follow up on its recommendations. But this did not happen, perhaps because others were becoming engaged. The National Science Advisor (Arthur Carty) was championing a federal S&T infrastructure fund in 2005, modelled on the Canada Foundation for Innovation.[47] However, the proposal was overtaken by other events and did not move forward. In the summer of 2006, TBS once again entered the scene seeking to follow up on the working group's efforts. TBS proposed a two-year undertaking, a Federal Laboratory Infrastructure Project. The first six months, which TBS was offering to fund, would be devoted to designing an infrastructure assessment process and a governance framework. This would be followed by a year of implementation and by final recommendations to DMs of science-based

---

[45] *Update on Capital Assets Review and Science & Technology Infrastructure*, presentation to EC on the report of the Science Infrastructure Review Working Group, October 20 2004. Science Policy Division files

[46] See *Laboratory Assets Review: The Federal Science Infrastructure Challenge* (April 2005) for a more detailed account of the working group's activities. Science Policy Division files

[47] The Canada Foundation for Innovation was created in 1997. By March 31 2009, it had invested $4.5 billion in infrastructure projects at universities, colleges, research hospitals and non-profit research institutions. *Canada Foundation for Innovation 2008-09 Annual Report*

departments and to the TBS Secretary. Departments would be asked to fund that phase. The project would be steered by its own ADM committee, involving both corporate and science ADMs, which would be co-chaired by TBS and the NRC. Departments agreed to be engaged in the first part of the project and to consider the rest later.

The first element of the project was to develop a rigorous process for assessing the physical conditions of S&T facilities and major equipment. The database of the earlier working group had been put together in a short time and suffered from definitional problems and inconsistencies in the data provided by departments. A consulting firm was engaged and an interdepartmental group set up to work with it. Together they scoped out the methodology for the inventory, defined key concepts and developed options on the level of detail that might be collected.

The second element was also handled by a consultant. Based on interviews with departments he prepared a discussion paper on the challenges and opportunities for a governance system for S&T infrastructure and suggested some areas for further study to flesh out the system.[48]

The ADM steering committee met on April 27, 2007 to review the design phase of the project and to decide on moving forward with the next phase, implementation. The committee was presented with three options for a standardized database for S&T infrastructure. These differed in coverage. For example, the modest option would include information on buildings worth more than $10 million (or of strategic value) and on equipment over $250,000; the comprehensive option would include buildings over $500,000 and equipment over $10,000. For EC, the first option would mean reporting on seven facilities and on 197 pieces of S&T equipment (excluding information technology); the second would include another 24 buildings and 5004 pieces of equipment.[49] In addition, the depth of information to be collected under each option varied.

The members of the committee had several concerns. EC, for example, did not think that the proposed building condition reports would provide enough information. It was also concerned about the costs of collecting the information – even the modest option would cost the Department about $500,000 and take 18 months to implement. The

---

[48] *Federal Laboratory Infrastructure Project (FLIP). Governance Report. Draft for discussion*, prepared by The Impact Group, March 2007. Science Policy Division files
[49] All but one of the facilities over $10 million were in the S&T Branch. The exception was a building shared with EC's Meteorological Service. In contrast, only 41 pieces of equipment over $250,000 were in the Branch with 153 in the Meteorological Service. *Federal Laboratory Infrastructure Project (FLIP): Scope and potential costs for Environment Canada.* EC presentation, 2007. Science Policy Division files

benefits of a common database were not well defined. TBS did not appear ready to provide any incentives; indeed, it was promoting other asset management initiatives in which EC was participating. And, the government had made a surprise announcement, the previous month, of an independent expert panel on the transfer of non-regulatory federal laboratories. This had the effect of postponing any further action on a common governance system at least until the panel had made its report.[50] At the conclusion of the ADMs meeting, it was clear that the Federal Laboratory Infrastructure Project had reached a dead end. The Project might have continued if either departments or TBS would have provided resources, but they did not.

In 2007, TBS was also promoting another infrastructure initiative, Integrated Investment Planning, at the time that the Federal Laboratory Infrastructure Project was being abandoned. The initiative was a consequence of two new TBS policies, one of which (investment planning for assets and acquired services) was a replacement for the policy on long-term capital planning. EC was part of the 2008-09 pilot for the initiative. The main difference for EC was that investment requirements now were prioritized departmentally rather than by program. The result in EC was the development of a stronger centralized capital management planning capacity. The broader impact of Integrated Investment Planning was to focus attention, when it came to S&T infrastructure, on departmental rather than interdepartmental planning.

The global financial crisis of 2008 led to the first separate set of funds for federal S&T infrastructure. The January 2009 Budget provided $250 million for deferred maintenance at federal laboratories and $85 million for maintenance or upgrading of existing Arctic research facilities, as part of the government's economic action plan.[51] The process was coordinated by TBS. EC was able to obtain over $14 million for investments at the National Wildlife Research Centre, the Environmental Science and Technology Centre, the National Water Research Institute, the Wastewater Technology Centre, a network of field stations for Arctic migratory bird research, and the Dr. Neil Trivett Global Atmosphere Watch Observatory in Alert, Nunavut.

---

[50] See chapter 8.
[51] $2 billion was also provided for maintenance and repair at post-secondary institutions.

Figure 5.3 Map of EC S&T Facilities, 2009

6

Managing Personnel

The scientists, engineers, technicians and technologists who work for Environment Canada are, without a doubt, the most important element of its S&T capacity. They have usually formed the majority of the Department's employees (see Appendix 2). Policy for S&T personnel covers a broad range of issues, including recruitment, compensation, retention, learning and employment equity. For the most part, the human resources issues faced by EC were not unique to it. They were common to most S&T-performing departments. That was because Treasury Board, the employer of all public servants, set out the key human resource policies of the federal government. On its own, EC had limited ability to amend them. However, working with other departments – especially when there were opportunities for change – boosted the chances of reshaping human resources policy.

EC's engagement in this policy area was intermittent up until the mid-1990s. Prior to that, the Department's efforts were largely responses to initiatives by central agencies. Within EC, senior science managers tended to provide policy leadership. The human resources group did not possess a strong policy capacity, being primarily occupied with administrative work. Nonetheless, it had an important role to play. Implementation of policies for S&T personnel was always much more complex and time consuming than their development. Successful implementation was usually the result of close collaboration at the development stage between science managers and human resources specialists.

In 1995, EC began a very active and extended involvement in S&T human resources policy. That year, Treasury Board established an interdepartmental ADMs committee to implement a response to a report by the Auditor General, which had raised a number of issues regarding the management of S&T personnel. The committee and its successors, which still continue to meet, undertook a considerable amount of policy work. In turn, the activity strongly influenced the agenda and pace of human resources policy within EC for its S&T workforce.

The interdepartmental committee had an important impact in fostering a sense of shared interest among senior science managers in departments that performed S&T. Its achievements gained the federal S&T workforce a reputation for excellence in community management. But by the mid-2000s, it was clear that this effort was increasingly at crosscurrents with public service reform. The Public Service Management Act of 2003 and the acts that followed it squarely placed responsibility for human resources management in the domain of individual departments. Collaborative, community-based policy effort on S&T personnel would play a secondary role as departments focused on taking up their new responsibilities.

Responding to S&T Personnel Issues

Engaging the Policy Area

EC did little policy work on its scientific and technical human resources in the 1970s. This was not because there were no issues. At the same time as EC was being put together, a review of the many and complex issues facing federal S&T employees was being prepared for the Gendron task force looking at science in the federal government.[1] Because the issues were common to many departments and were usually seen inside the government as subordinate to broader policy for civil servants, EC likely looked to central agencies – in particular Treasury Board, the Public Service Commission and MOSST – for leadership. The work the Department did perform in those years was usually in response to studies conducted by these agencies.[2]

In its second decade, EC began to be more proactive in its attention to S&T personnel. The topic was one of five areas covered by the 1980 comprehensive review of science management in the Department, accounting for about a third of all the follow-up actions.[3] But it was still central agencies who were most engaged by the issue. Prompted by dissatisfaction among employees due to tight S&T expenditures and the government's lack of interest in its own S&T effort, they undertook several studies. An interdepartmental working group on S&T human resource management was formed in 1986, co-chaired by MOSST and the Treasury Board Secretariat (TBS). The following year, the Professional Institute of the Public Service of Canada, the union representing most of the scientists and engineers in the federal government, presented a report to the group laying out six major problems and suggesting possible solutions.[4] The working group issued its own report in 1992, identifying four issues requiring attention: revitalization, motivation and morale, management skills, and the under-

---

[1] W. L. Ellis, *A Study of Scientific Personnel in Government*, December 1970. I have not seen a copy of this report.
[2] See the Key Issues section of this chapter for examples of this work.
[3] See Key Issues later in this chapter.
[4] *The Management of Science and Technology in the Public Service: The Scientist's Perspective*, 1987

representation of women.[5] Two years earlier, the Prime Minister's National Advisory Board on S&T had called for a revitalization of the federal S&T workforce and noted a number of shortcomings in the management of S&T staff, including a need for rejuvenation, new rewards practices, more attention to management development, and an effective promotion process.[6] In 1994, the Professional Institute released a paper disclosing the results of an opinion survey of its members.[7] Government scientists and engineers perceived an out-dated, non-productive command and control culture, negative attitudes to professional staff, suspect staffing practices, limited training and career development, and lack of progressive leadership by middle and senior management as key issues requiring attention.

As part of its major review of federal S&T in 1994, the Auditor General devoted one of the chapters of its report to the management of scientific personnel in federal research establishments.[8] Its focus was on two issues: attention to long-term human resource requirements and development of an effective research management capability. The Auditor General made several recommendations. Science-based departments were advised to develop strategies for the management of their scientific personnel. Treasury Board was urged to review its human resource policies and practices to provide greater flexibility to those departments. And the government was asked to create a broad forum, drawing upon S&T managers in and outside the public service, dedicated to federal S&T personnel issues.

The Auditor General noted that his report was not the first study of federal S&T to raise concerns about the government's S&T workforce and that many of the problems that had been identified had not yet been resolved. The causes of this inertia were diagnosed as the failure to assign responsibility for implementing the recommendations, uncertainty about future S&T activities and budgets, and systemic constraints to effective management.

[5] *Strategic Issues, Objectives and Recommendations for the Management of the Scientific, Engineering and Technical Community in Key Federal Science Based Departments*, prepared by the Working Group on the Management of S&T Human Resources, November 1992
[6] *Revitalizing Science and Technology in the Government of Canada*. Report of the Committee on Federal Science and Technology Expenditures. National Advisory Board on Science and Technology, 1990. This is popularly known as the Lortie report.
[7] *Issues Facing Professional Employees in the Federal Public Service*, The Professional Institute of the Public Service of Canada, February 1994
[8] Chapter 11

TBS S&T Steering Committee

For a while it seemed as if the Auditor General's report would meet the same fate as earlier studies. However, the House of Commons Standing Committee on Public Accounts held a number of public hearings on the report the following year, in the spring of 1995.[9] In its findings, the Committee asked "whether the approach taken to the management of scientific human resources in the Public Service should not be changed" and noted that the Treasury Board, as the employer of the public service, had an important role to play in this.

TBS took up the Standing Committee's challenge. Staff from its Human Resources Branch consulted with government scientists, union representatives and ADMs of the major science-based departments. They then drafted a "Framework for the Human Resources Management of the Federal Science and Technology Community" and circulated it for comment. It laid out the S&T personnel challenges faced by departments and proposed five projects for addressing them. An interdepartmental Steering Committee was established to oversee the projects and implement the Framework (Figure 6.1). In addition, a small S&T unit was created at TBS to support the work of the committee.

The Steering Committee was chaired by TBS and composed of ADMs or their equivalents from the major science-based departments and agencies and from the Public Service Commission. In an unusual move – intended by TBS to signal a new way of doing business – the committee also included a union head, the President of the Professional Institute of the Public Service of Canada.[10] The first meeting of the steering committee was held on November 17, 1995.

Work on the five projects – management and scientific development and training, classification and compression, workforce and mobility, recruitment and rejuvenation, and rewards, recognition and incentives – commenced soon afterwards. A sixth working group, focused on the issues of technicians and technologists, was created in October 1996.[11] A different department led each project, but the work was planned and carried out through interdepartmental working groups. Over 100

---

[9] *Sixteenth Report*, Standing Committee on Public Accounts, 1995

[10] The Public Service Alliance of Canada was also approached but declined to participate. That union represents, within the federal S&T workforce, federal technicians and technologists.

[11] The workforce and mobility team was originally intended to deal with this topic. But it was decided that a separate team was needed. In addition to departmental representatives, the Public Service Alliance of Canada, the International Brotherhood of Electrical Workers and the National Research Council Employees Association participated on the working team. EC Science Policy Branch, box 40

scientists, managers and human resources specialists were involved. EC led the workforce and mobility project and had several employees on each of the working groups. In addition, the Department was responsible for four of the ten pilot projects run under the Framework.[12]   EC's participants also met together during 1996 to compare notes and to ensure that there was a good flow of information between the exercise and the Department, in particular its human resources directorate.

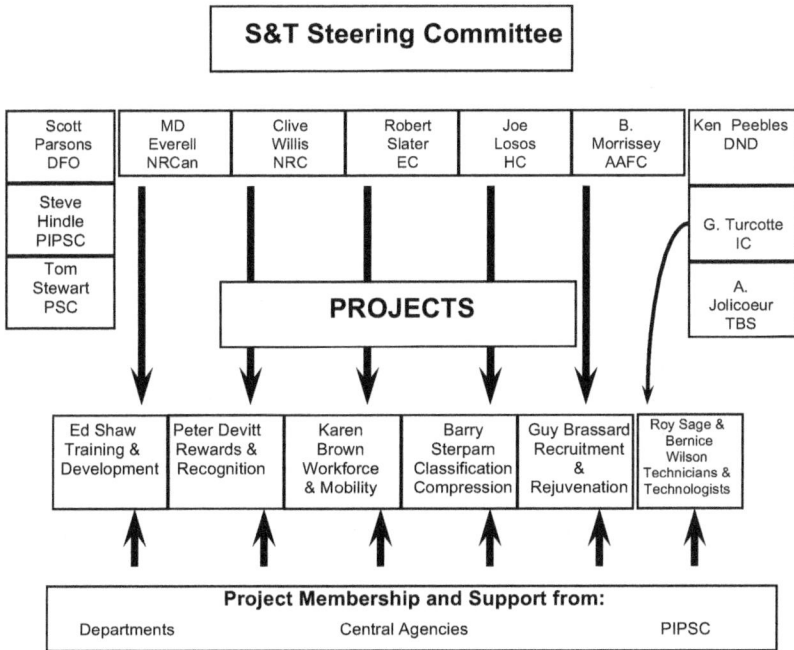

S&T Steering Committee

Scott Parsons DFO	MD Everell NRCan	Clive Willis NRC	Robert Slater EC	Joe Losos HC	B. Morrissey AAFC	Ken Peebles DND

| Steve Hindle PIPSC | | | | | | G. Turcotte IC |
| Tom Stewart PSC | | PROJECTS | | | | A. Jolicoeur TBS |

Ed Shaw Training & Development	Peter Devitt Rewards & Recognition	Karen Brown Workforce & Mobility	Barry Sterpam Classification Compression	Guy Brassard Recruitment & Rejuvenation	Roy Sage & Bernice Wilson Technicians & Technologists

Project Membership and Support from:		
Departments	Central Agencies	PIPSC

Figure 6.1 S&T Human Resources Framework
Steering Committee members, projects and project leads.[13]

Just prior to the Framework exercise, the scope of EC's policy work on its S&T workforce was fairly narrow, focused on dealing with the staffing cuts resulting from Program Review and on ensuring it would

---

[12] The four were the Green Corps, NRWI. com, People & Jobs, and a Competency Assessment Tool Kit and Personal Development Planner/Guide for NWRI. For information on these and the other pilots see *Assessment of the Science and Technology Human Resource Pilot Projects. Draft Report*, Consulting and Audit Canada, March 1998.

[13] This diagram, dating from March 1997, also shows the additional, sixth project on technicians and technologists which was added in 1996.

have the S&T skills it would need in the future.[14] The Framework broadened that scope and became the basis for human resource planning and reporting on the Department's scientists, engineers, technicians and technologists. The Department was very actively engaged in the Framework effort. EC's leads on the exercise made regular presentations back to the departmental management committee as well as to other departmental committees. When the Clerk of the Privy Council called for departmental human resource action plans, EC's DM supported the work under the Framework in his response.[15] The Department's plan was strongly influenced by the Framework.

The work under the Steering Committee was soon annexed to the release of the new federal S&T strategy in the spring of 1996. The Committee's Framework was published as part of a series of action plans that accompanied the strategy. A joint communiqué from the Presidents of the Treasury Board and the Professional Institute of the Public Service was also produced to increase awareness of the Framework initiative and to show that management and union were working together. However, the work on the Framework was completely separate from that done on the strategy or under the interdepartmental ADMs Committee on S&T. The Steering Committee provided a forum where senior S&T managers and policy staff from federal science-performing departments got to exchange views, work together and better know one another. Along with the fora provided by the *Memorandum of Understanding between the Four Natural Resource Departments on Science and Technology for Sustainable Development* and the ad hoc ADMs committee on science in government, the Steering Committee served as an incubator for a sense of community and common purpose among science-performing departments.[16]

During its first year, the Steering Committee's attention was focused on the progress of the projects. By the end of the year several projects were well advanced and had draft recommendations. The Auditor General's report back to the Standing Committee in June 1996 was positive about the Framework, describing it as a major step forward, although it also noted that much remained to be done.[17] This assessment

---

[14] *An Action Plan for Managing Science and Technology at Environment Canada*, November 1994. EC OSA, boxes 4&7

[15] *Human Resources Management Plan 1997/98 – 2001/02. Situating Environment Canada for the Future: An Integrated Human Resources Action Plan*, presentation to Committee of Senior Officials (COSO), May 1997. EC Science Policy Branch, box 42

[16] See chapter 7 for further information on the Memorandum, and chapter 10 for the ad hoc committee.

[17] Letter from the Auditor General of Canada to the Chairman, Standing Committee on Public Accounts, June 7 1996. EC Science Policy Branch, box 37

was repeated in the Auditor General's September report, with an emphasis on the need to pay attention to implementation.

> Although we are encouraged by the progress reflected to date in the *Strategy* and the *Framework*, implementation is the real challenge. For this reason, we believe that the government needs to devote considerable attention to establishing results-oriented, time-phased implementation plans for both the *Strategy* and the *Framework*. The government also needs to ensure that accountability for results is clearly established. Implementation will require leadership and perseverance at all levels of government – from ministers to scientists. Parliamentary oversight also needs to continue.[18]

The Steering Committee was also concerned about implementation. For example, although departments were responsible for implementing the Committee's findings, members believed that departments would have a hard time dealing with the separate reports they were producing, the large number of recommendations (there would be over 140 from the exercise) and the multiple suggested timeframes. The Committee struck a working group in the fall of 1996 to develop an implementation strategy and then carry it out.

The working group decided that it needed to consolidate the recommendations that were beginning to emerge from the projects. The result was a table spread over more than 20 pages named the *Blueprint for Action*. It was organized under four themes: rejuvenating the science workforce, new work environment, investing in people, and governance. For each recommendation, the Blueprint provided notes on deliverables, implementation activities, projects and pilots, roles and responsibilities, timing, resources and costs, and do-ability. Updates of the Blueprint were produced as new recommendations emerged. In association with the Blueprint, the working group also produced an accountability framework. It listed the deliverables, their critical success factors, performance indicators and possible sources of information.

Another implementation step was a series of consultations carried out across the country in February and March 1997.[19] It originated in the interest shown by three of the projects in conducting consultations. The working group decided to have one set of consultations integrating the

---

[18] "Federal Science and Technology Activities: Follow-up," chapter 15 in the *1996 Report of the Auditor General of Canada to the House of Commons*. The *Strategy* referred to in the quote is the 1996 federal S&T strategy.
[19] *Consultation Data Reports*, prepared by Consulting and Audit Canada, March 1997. EC Science Policy Branch, box 38

needs of the projects. It was also seen as an opportunity to test the recommendations and communicate the Framework. The consultations were funded by the TBS. Over 300 government S&T employees participated. In addition, a series of focus groups with students were conducted in April to determine what the government could do to attract and recruit them.[20] These involved 110 students. The results of the consultations helped to validate or refine the projects' recommendations.

Although it was clear that the success of the Framework exercise ultimately depended on the implementation of its work by departments, the members of the Steering Committee believed that cross-departmental leadership had a key role to play.[21] An opportunity to pursue this direction emerged in early 1997, with a new high-level initiative to rejuvenate the public service.

S&T Community Management

The Clerk of the Privy Council launched La Relève in her *Fourth Annual Report on the Public Service of Canada* in February 1997. It was a call to action to build a modern and vibrant public service. One of the elements of the Steering Committee's implementation strategy had been to get on the agenda of La Relève. It was completely successful in that regard. The work of the Committee was mentioned in the *Fourth Annual Report*. A DM champion for S&T was named.[22] A DM-level committee on S&T was also created, having its first meeting in March.[23] And the Steering Committee presented its work plan to the Clerk as part of La Relève, in April.[24] The Framework exercise had found new support, moving from a special initiative responding to the concerns of the Auditor General to being part of a broad renewal of human resource management practices in the government. In the process, influenced by La Relève's support for "functional communities" within the public service, the Framework exercise began to brand itself as S&T community management.

---

[20] *Review of Student Focus Groups and Student Fact Sheets*, April 1997. EC Science Policy Branch, box 38

[21] Minutes of the meeting of S&T Team Leaders Plus, October 21, 1996. EC Science Policy Branch, box 38

[22] The DM champions were William Rowat (1997), Jean McCloskey (1997-1999), Frank Claydon (1999-2000), and Peter Harrison (2000-2003).

[23] This was a subcommittee of COSO, the Committee of Senior Officials that advised the Clerk. The subcommittee would meet, with a two-year hiatus from 1999 to 2001, until the spring of 2003. EC Science Policy Branch, box 37

[24] *La Relève: Commitment to Action. Functional Plan. Science and Technology Community*, April 1997. EC Science Policy Branch, box 44

Another mark of the evolution towards community management was the decision by TBS that the Steering Committee should have a co-chair from one of the science-based departments. Reflecting the effort that EC had put into the Framework as well as her own abilities and interest in the subject, Karen Brown (then EC's ADM on the Steering Committee) was chosen to be co-chair.[25]

The nature of the work under the Framework also began to shift. The Framework's projects had mostly come to an end by the close of 1997. The two years since their start up had been devoted to carrying out broad-ranging studies in the six theme areas, and then to consolidating the resulting recommendations, to consultations and to oversight of a number of pilot projects. Spurred on by the DM champion at that time (Jean McCloskey), the Steering Committee began to tackle specific issues with the intention of taking action to resolve them (some of this work is reviewed later in this chapter).[26]

A clear indicator of this shift was the February 1998 Steering Committee retreat at which the DM champion participated. Its purpose was to reflect on the Committee's "activities to date, to discuss difficulties and barriers, and to develop priorities and action items for 1998-99."[27] The Committee was challenged by the DM to "focus on a few key areas and to demonstrate clear achievements." In response, the members decided to direct their efforts to assisting with the application of the Universal Classification Standard to federal S&T jobs, to completing the work of the technicians and technologists team, and to improving their communications with the federal S&T community.[28] They also reaffirmed the three new working groups – on a demographic analysis of the community, on developing a competency profile for S&T managers, and on the promotion criteria for research scientists – which had been initiated after a discussion of the Framework's work plan by the DM committee the previous fall. And they established a new working group, to be led by EC, to review issues facing women in federal S&T.

These areas were the primary focus for the Steering Committee over the next few years. The emphasis continued to be on taking action. For example, TBS was convinced in 1999 to resolve two issues that had long been major irritants for scientific researchers. TBS clarified the policy on conference attendance by noting that conferences should be considered

[25] Karen Brown became co-chair after the February 1998 retreat, serving until 2005.
[26] A detailed summary of the work of the Steering Committee can be found in *A Review of Federal Science & Technology Human Resources Studies, 1994-2001*, report prepared by Ruth Matte Consulting for the Council of Science and Technology Advisors, February 2002.
[27] *Science & Technology Human Resource Steering Committee. Update on Implementation & Next Steps*, 1998. EC Science Policy Branch, box 40
[28] For more on the Universal Classification Standard, see later in this chapter.

as training for scientists, thereby removing a major barrier to their attendance. And TBS lifted the quotas on the number of research scientists allowed within each level of their classification.[29] The Committee also followed up on its commitment to increase communications with the federal S&T community. In December 1998 the first federal science managers' forum was held. It attracted about 200 participants, including 21 from EC. The forum's purpose was to help disseminate the work of the Committee to the broader federal community. In the same vein, the forum was followed up by six regional fora between 2000 and 2002, which were organized in concert with the Regional Federal Councils.

At this time there was some talk about winding up the work of the Steering Committee. The Auditor General was satisfied with what had been done.

> Overall we think that the Treasury Board Secretariat, science-based departments and agencies and the science and technology community have shown leadership in the management of their human resources. Based on the strength of the community's achievements and progress despite the difficult environment, some senior officials have suggested that the science and technology community could serve as a model across the federal government.[30]

However, managing S&T personnel was not an issue that could simply be solved and forgotten about. It was a perennial issue, ready to respond to changes in its environment – one which was quite active at that time. Even as the Auditor General's report was acknowledging the Committee's efforts, it noted that "there are still major challenges to be met," mentioning the worsening demographic profile of the federal S&T community. And many other opportunities for engaging in work on federal human resources management were emerging. Among these were the follow-up to the abandoned work on the Universal Classification Standard, the formation in 2001 of the Task Force on Modernizing Human Resources Management in the Public Service, and the study by the Council of Science and Technology Advisors of the key S&T human resources challenges facing the federal government in the 21st century.[31]

---

[29] See later in this chapter for more on this.
[30] "Management of Science and Technology Personnel: Follow-up," chapter 9 in the *1999 Report of the Auditor General of Canada to the House of Commons*
[31] *Employees Driving Government Excellence (EDGE): Renewing S&T Human Resources in the Federal Public Service*, November 2002

Despite the ongoing interest, TBS began to signal that it would be withdrawing its support from the Steering Committee. This was a retreat from earlier funding for functional communities. It revealed a change in attitude towards programs that cut across departmental lines, as well as an inability to deal effectively with horizontal issues.

A good example of the change in relationship is the experience with the Graduate Opportunities Strategy initiative. It began to be developed by the Steering Committee in 1999 in response to demographic concerns and to an emphasis on recruitment by the Clerk.[32] It proposed a bridging fund to hire new, replacement S&T staff in advance of the retirement of existing staff to help ensure that the latter's expertise would not be lost. In addition, an emphasis was placed on increasing the representativeness of employment equity groups (women, aboriginals, the disabled and visible minorities) through the hiring. The original proposal was for $200 million over five years, covering 12 departments and agencies. TBS was not comfortable with such a large amount, even though it covered the large majority of the government's S&T workforce. After considerable scenario work and discussions, the initiative eventually received $3.75 million as a one-year pilot covering only research scientists and some technicians (the EG classification – the engineering and scientific support group). EC received $428,000 from this amount, enough to hire eight individuals. Although the program was judged a success, it was not repeated or further developed. The Steering Committee had to look for other ways of addressing recruitment and employment equity issues.

By the spring of 2001, the science ADMs on the Steering Committee had decided that their collective effort should continue, even without the assistance of TBS.[33] The Committee would no longer be co-chaired by TBS, although the latter would still be a member. Its secretariat would move from TBS to the department of the DM champion for S&T (at that time Peter Harrison of Natural Resources Canada). Most significantly, the eight science-based departments and agencies would now share in the funding needed for the secretariat and the Committee's activities. Each agreed to provide about $58,000 annually for the three-year period 2002-03 to 2004-05, with TBS providing $500,000 for the 2001-02 transition year. To signal the break, the Committee changed its name to the Science ADMs Advisory Committee on Human Resources.

The activities pursued by the Advisory Committee continued as before. The emphasis was on specific action items within the themes of

32 *The Graduate Opportunities Strategy (GOS): Preliminary Assessment and Suggested Evaluation Framework*, Consulting and Audit Canada, March 2002
33 *S&T ADM Senior Steering Committee – Future Directions*, report for the Committee, May 2001. EC Science Policy Branch, box 39

recruitment, retention and learning. The Committee focused on topics where collective action could be taken and which would be of benefit to the whole federal S&T community. It continued its communications efforts. And it represented the perspective of its community in relevant government initiatives, for example submitting recommendations to the Task Force on Modernizing Human Resources Management in the Public Service.[34] The Committee also prepared a five-year strategy to guide its work, as well as annual work plans and reports.[35]

In 2004, the Advisory Committee's secretariat began to house the Integration Board's one-person support staff.[36] Since the membership of these two ADM-level committees was almost the same, the two groups were merged in 2007. And as the science ADMs were mostly preoccupied with the issue of integration at that time, the resulting committee continued to use the Integration Board name. Despite the disappearance of an ADM committee solely focused on S&T human resource issues, the new Integration Board has continued to devote a considerable amount of its time to this area of policy for S&T.

Key Issues

Policy regarding S&T personnel covered a variety of issues and involved many activities, especially after the formation in 1995 of the ADMs' Steering Committee for the S&T Human Resources Framework. This section reviews four issues – classification, promotion of research scientists, development of S&T managers, and recruitment and employment equity. These were the subjects of most of EC's policy effort on S&T personnel.

Classification

All federal jobs are classified into one of over 60 occupational groups. Each group has a classification standard detailing the work expectations for every level within the group. About a third of the classifications cover scientific and technical occupations. The standards provide the basis for determining the level of a specific job and hence its remuneration. That is why classification can be the subject of controversy, especially as jobs evolve over time and issues such as pay equity come into play.

---

[34] August 2001. Science Policy Division files
[35] *The Science and Technology Strategy for Effective Human Resources Management, 2004-2009*, 2004. Science Policy Division files
[36] See chapter 7 for a discussion of the Integration Board.

During its first twenty years, EC policy for science did not appear to be much concerned with classification issues. An exception was at the time of the 1980 Science Review. One of its follow-up actions was to gather together the Department's concerns about the classification of technical staff in engineering and scientific support, the single largest group of S&T employees in EC.[37] This was done as part of a major review of that group's classification standard by Treasury Board.

The Department's policy involvement in classification greatly increased with the S&T Human Resources Framework. One of the latter's projects was "classification and compression." The government had been looking to reform classification and reduce the number of its occupational groups since at least 1990.[38] At the same time as the Framework was being established in 1995, TBS was embarking on a second attempt at dealing with the issue, through usage of a Universal Classification Standard throughout the government. It was an attempt to have only one standard for all federal jobs.

The Framework's Steering Committee took the Universal Classification Standard initiative very seriously, putting considerable effort and resources into it over a four-year period. Indeed, the S&T community probably provided more input into the Standard than any other group in the federal government. One of the first activities of the interdepartmental classification and compression project team was to define, in February 1996, the federal S&T community.[39] It decided to take a broad approach, including scientists, engineers, technicians and technologists who were not involved in research. The result of the team's work, with some slight modifications, became the standard one for the community (see Figure 6.2 for an example).

The project team also looked at group restructuring. It recommended one group with three sub-groups (research, applied science and engineering, and health sciences). It believed that a single group would help to increase mobility, consistency of salaries among departments, and teamwork as well as creating a stronger and more integrated S&T community. This recommendation was reviewed at an EC science forum in October 1996 and endorsed. Nonetheless, in the end TBS established three groups for scientists and engineers – research, applied science and patent examination, and health services – with other groups for technicians and technologists.

---

[37] See Figure 6.4
[38] *Public Service 2000: The Renewal of the Public Service of Canada*, November 1990
[39] *Report on the Definition of the Science Community*, February 13 1996. EC Science Policy Branch, box 42

Assessing whether the Universal Classification Standard could measure all of the S&T work performed by the government would be a long and complex task. In the first place, the S&T community was among the largest groups of employees within the federal government, at over

Occupational Category	Occupational Group
Applied Science and Engineering	Agriculture (AG)
	Biological Sciences (BI)
	Chemistry (CH)
	Engineering (EN-ENG)
	Land Survey (EN-SUR)
	Forestry (FO)
	Meteorology (MT)
	Physical Sciences (PC)
	Scientific Regulation (SG-SRE)
	Patent (SG-PAT)
Health	Medicine (MD)
	Nursing (NU)
	Pharmacy (PH)
	Veterinary Medicine (VM)
Technical	Engineering and Scientific Support (EG)
	Electronics (EL)
	General Technical (GT)
	Drafting and Illustration (DD)
Research	Scientific Research (SE)
	– Research Manager (REM)
	– Research Scientist (RES)
	Defence Scientific Service (DS)

National Research Council S&T Occupational Categories

Occupational Category	Occupational Group
Research	Research Officer (RO)
	Research Council Officer (RCO)
Technical	Technical Officer (TO)

Figure 6.2 List of S&T occupational categories and groups used by federal departments in 2002.[40]

[40] Taken from *Employees Driving Government Excellence (EDGE): Renewing S&T Human Resources in the Federal Public Service*, November 2002

20,000 individuals (Figure 6.3). EC was home to about 2700 of this population (Figure 6.4), making it one of the largest departmental employers of S&T personnel. This number was about 60% of the Department's total employees at that time. S&T staff was the majority community in EC.

Secondly, compared to some other communities, such as finance or human resources, the federal S&T community performed a wide variety of different functions. All of these would need to be evaluated. Despite the challenges, the project team had early in 1996 judged that the Universal Classification Standard could deal with the S&T community, but not without several revisions. In particular, the special circumstances of researchers needed to be addressed (see next section).

Functional Area	Occupational Group	Population 03/31/99	Percent of Functional Area	Percent of Total S&T Community
Applied Science	AG Agriculture	171	2.9%	0.9%
	AR Architecture and Town Planning	43	0.7%	0.2%
	BI Biological Sciences	1,581	26.9%	7.9%
	CH Chemistry	420	7.2%	2.1%
	EN Engineering and Land Survey	1,446	24.6%	7.2%
	FO Forestry	80	1.4%	0.4%
	MT Meteorology	485	8.3%	2.4%
	PC Physical Sciences	1,262	21.5%	6.3%
	SG Scientific Regulation & Patent	382	6.5%	1.9%
Applied Science Total		**5,870**	**100.0%**	**29.3%**
Health	MD Medicine	107	8.7%	0.5%
	NU Nursing	603	49.0%	3.0%
	PH Pharmacy	8	0.7%	0.0%
	VM Veterinary Medicine	512	41.6%	2.6%
Health Total		**1,230**	**100.0%**	**6.1%**
Research	DS Defense Scientific Service	437	19.7%	2.2%
	SE Scientific Research	1,782	80.3%	8.9%
Research Total		**2,219**	**100.0%**	**11.1%**
Technical	DD Drafting and Illustration	284	2.6%	1.4%
	EG Engineering and Scientific Support	4,845	45.2%	24.2%
	EL Electronics	995	9.3%	5.0%
	GT General Technical	1,758	16.4%	8.8%
	PI Primary Products Inspection	1,772	16.5%	8.8%
	PY Photography	28	0.3%	0.1%
	TI Technical Inspection	1,043	9.7%	5.2%
Technical Total		**10,725**	**100.0%**	**53.5%**
Grand Total		**20,044**		**100.0%**

**Figure 6.3 Size of the federal S&T community.
Prepared by the Research Directorate,
Public Service Commission of Canada, February 7, 2000**

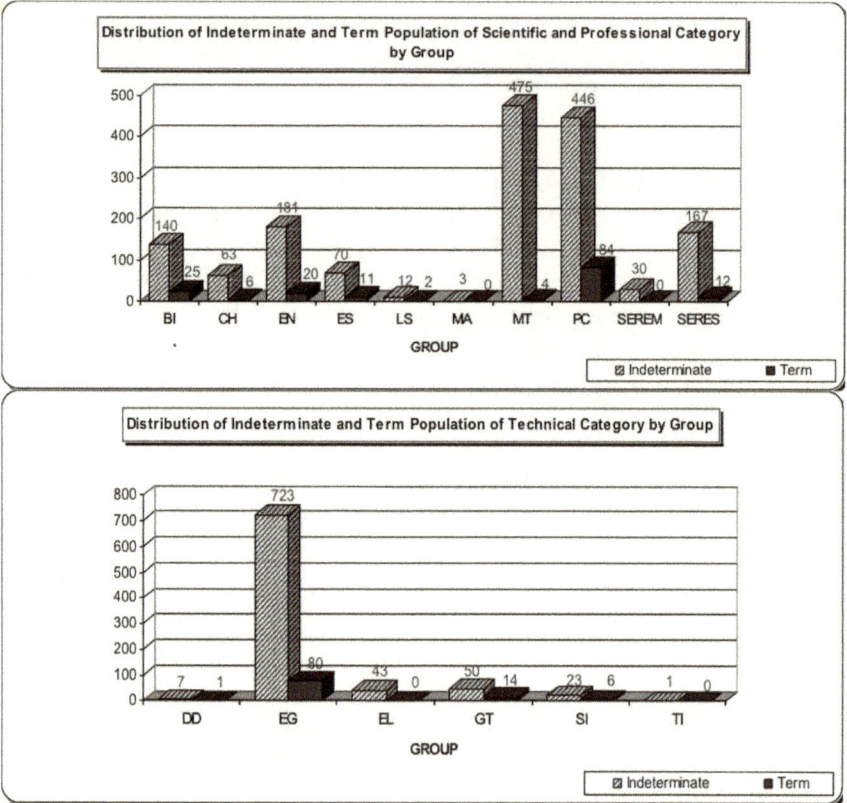

Distribution of Indeterminate and Term Population of Scientific and Professional Category by Group

Distribution of Indeterminate and Term Population of Technical Category by Group

Figure 6.4 Distribution of S&T occupational groups in EC, July 1998.[41]

For the next three years, the project team and other task groups set up by the Steering Committee provided advice to the designers of the Standard, helped develop model work descriptions and shared information across departments. They were particularly active in 1998, when TBS announced its objective to have a job description prepared for every federal employee based on the Standard. This caused a very large effort within departments, including EC. The work under the Steering Committee helped to ensure consistency of application of the Standard across departments.

---

[41] *EC Overview of Workforce*, EC S&T Advisory Board, box 5. Indeterminate signifies permanent employees; Term employees are those hired for specified periods.

As more draft Standard work descriptions became available, other issues came to the fore, such as the number of levels within the Standard and pay bands. It became clear to TBS in early 2001 that the move toward a government-wide Universal Classification Standard would not work. The following year, an official announcement was made that the Standard would not be implemented.

Promotion of Research Scientists

Unlike almost all other occupational groups in the federal government, the research scientist classification (SE-RES) allows its members to be promoted according to their individual productivity rather than by winning a competition for another position. In principle, all research scientists could get annual promotions. Attempts to control the rate of promotion as well as efforts to give greater weight to the relevancy of researchers' work provided the basis for discord.

When EC was established in 1971, the scientific research classification was five years old.[42] At that time the classification had four levels, with promotion a combination of experience and productivity.[43] The Department had 590 research scientists, about 8% of its S&T workforce.[44] Of the total, 153 researchers were at level one (the entry level), 366 at level two, 57 at three and 14 at four. While they represented a small segment of EC's staff, they were seen as critically important to the vitality and effectiveness of the Department's research effort.

The use of quotas was an early policy issue in the promotion of researchers. In 1974, quotas were established for levels three (at 25% of the total) and four (5%). In addition, an interdepartmental advisory committee was established to monitor promotions. This practice caused administrative difficulties for departments. For example, when the Canadian Forestry Service was moved out of EC in 1984, the distribution of the remaining researchers was not aligned with the quotas. Given the circumstances, EC was granted a five-year exemption. But in 1990 the

---

[42] Robin Reenstra-Bryant, *The Promotion System for the Research Scientist (SE/RES) Classification*, MOSST, January 1984. EC Roots, box 6. See also Les Carlson, *Historical Classification Background: Scientific Research*, November 1994.
[43] Ibid.
[44] By Service, the Atmospheric Environment Service employed 40 SE-RESs; Fisheries 166; Lands, Forests & Wildlife 281; Water Management 94; and Finance and Administration 3. Peter Meyboom, *Science in a Changing Environment – proposals for a departmental science policy*, 1972

Department was still not compliant, having 6% of its research scientists at level four and 30% at three.[45]

By the late 1980s, pressure was building to amend the levels due the lengthy period many researchers were spending at level two. TBS revised the classification standard in 1990, introducing a new level between the previous two and three. The quotas were amended so that the new highest level (level five) was restricted to 5%, the next to 25% and level three to 20%.

The system led to much discontent among researchers who complained about the inconsistency of using quotas in a practice that was intended to promote based on individual achievement. A report to TBS in 1997 noted the unease and recommended eliminating the quotas, which TBS did in 1999.[46] Their removal did not result in the fear some had that research scientists would cluster at the top level. EC, for example, had 206 individuals in the scientific research classification in 2008. Of these, 9% were at the highest level, 21% at level four, 34% at three, 30% at two, and 6% at level one.[47]

A second issue was the procedure used by EC to promote researchers. The 1980 Science Review had flagged the appraisal and promotion process as a concern. As a result, the ADM of Finance and Administration and DG of Personnel studied the question, identifying a number of problems and making several recommendations.[48] The departmental management committee then asked the Science Advisor in March 1982 to chair an ADM committee to develop new procedures. This was quickly done, the new process being approved in November.[49] The major changes were boosting the membership of the departmental promotion committee from mid-level officials to senior level (the chair was now an ADM), including two senior researchers on the committee, and making the promotion committees in the Services consistent. This system has served the Department well and is still the one currently used.

The third issue was keeping the criteria used to assess researchers' applications for promotion up to date. The 1997 report to TBS noted that departments wanted to update the criteria. Later that year, the Steering Committee created a special working group to look at the

[45] *Options for the Management of the DOE Research Scientist Population Distribution Dilemma*, EC Science Committee meeting, January 1990. EC S&T Management Committee, box 1
[46] *Research Scientist Quotas. Draft report on the consultation with S&T departments*, prepared by Price Waterhouse for TBS, 1997. EC Science Policy Branch, box 43
[47] *Analysis of RES Promotions, 1993-2008*. Science Policy Division, October 2008
[48] Meeting, March 17 1982. LAC Acc. 1991-92/017, box 38. See also Jocelyn L. Veilleux, *Système de Gestion du Personnel des Chercheurs Scientifiques à Environnement Canada*, course paper at École nationale d'administration publique, February 1982. EC Roots, box 9
[49] Meeting, November 24 1982. LAC Acc. 1991-92/017, box 39

promotion of research scientists. It was supposed to deal with both the criteria and with the question of whether the practice of promoting researchers was compatible with the Universal Classification Standard. However, the latter issue had priority. The promotion criteria would have to wait.

The attention of the working group was focused on the Standard for the next two years. The research scientist category had been placed into a Research Group that also contained the classifications for defence scientists, mathematicians and historians, each of which also had elements of promotion based on achievement. A great deal of the working group's time was devoted to an examination of whether one system could be used for all these classifications. The rest of the time was spent in examining the implications of the Standard for the existing promotion systems. The core difficulty was that the Standard was a position-based system, while they were using individual-based ones. So the central question was how an individual-based system could work under the Standard. There were also other questions: could the Standard capture research jobs, how many levels of research scientists could be distinguished under the Standard, and would positions in other categories migrate into the Research Group.

With the demise of the Universal Classification Standard initiative, TBS had decided to move forward by working on individual standards for selected groups. The ADMs Steering Committee successfully lobbied to have the Research Group included in the initial selection. A new working group, the Research Community Advisory Committee, was established in September 2000. It was chaired by TBS and was intended to be an employer-union body, although it also included researchers and managers from science-based departments (EC was actively engaged). It was this Committee that produced in 2005 a new framework to guide the promotion of scientific researchers.[50] EC adopted it immediately and has used it since the 2006-07 fiscal year.

Development of S&T Managers

Science managers have long been recognized as key figures in ensuring the quality and relevance of federal S&T. As a consequence, their development has also long been the subject of policy work.

---

[50] *Career Progression Management Framework for Federal Researchers: Application for the SE-RES Community*, February 2006

The Public Service Commission conducted a study of the training needs of research managers in 1972.[51] It followed this up in 1974 with another report that included several recommendations on the subject.[52] It urged a "unified and coordinated approach to the training and development of research managers" as well as an expansion of the Commission's course on research management. The result was a three-week course in the Commission's program of executive development seminars. EC participated in the studies by surveying its S&T staff for their views on the obstacles to management development.[53] It found that, while scientists viewed management as a valid career goal and the identification of potential managers was not a problem, much more use could be made of development leave for research scientists. The report was widely distributed in the Department and accepted in principle by the departmental management committee.[54]

MOSST was also involved in the issue of developing science managers. It produced a paper in 1979 on the topic as part of a larger study on federal scientific personnel.[55] It claimed that science managers required special training:

> ... to be an effective science manager, knowledge of general management is necessary but not sufficient. Special skills and attributes are needed to manage science activities and highly qualified personnel. An appropriate mix of talents and skills will have to be achieved comprising the specialized knowledge and experience for research management with the aptitude and skills for general management.

The paper recommended that science-based departments should consider developing modules for scientists in their training programs.

Concern about the skills of science managers was also strong in EC at that time. It was one of the items covered in the 1980 Science Review. In preparatory work, EC had surveyed 26 research managers in December

---

[51] *Training Needs of Research Managers*, Public Service Commission, Bureau of Staff Development and Training, August 1972. Cited in *Scientific Manpower in the Federal Government (Phase II). Background Papers*, MOSST, April 1979.

[52] *Public Service Manpower Survey: Scientific Research Group*, Public Service Commission, Manpower Planning Division, March 1974. Cited in *Scientific Manpower in the Federal Government (Phase II). Background Papers*, MOSST, April 1979.

[53] *Report on the Development of Scientific and Professional Staff for Management*, 1975. Discussed in LAC Acc. 1993-94/003, boxes 2 & 20.

[54] Meeting, November 20 1975. LAC Acc. 1991-92/017, box 15

[55] "Training and Development of Research Managers in the Public Service," *Scientific Manpower in the Federal Government (Phase II). Background Papers*, MOSST, April 1979

1979 about their training needs.[56] An outline for a three-day course was developed, but not implemented. Other initiatives, such as training and development plans for all employees, were given priority. Still, S&T staff did receive at least general management training. In 1983-84, for example, about 280 underwent supervisory training and over 100 middle management training.

The challenge of developing science managers came to the fore again in 1995 as one of the core issues of the TBS Framework for the management of S&T human resources. The challenge was identified as increasing the number of scientists moving into the management or executive category. Under the influence of La Relève, as mentioned before, the work of the Framework's ADMs' Steering Committee had become much more targeted and action-oriented. The issue of developing science managers provides a good example of the shift. One of the original project's recommendations had been to define the learning needs of science managers.[57] Now, in response to a presentation to the DMs committee in 1997, a new Scientific Management Development working group was established. Its goal was to follow up on the project's recommendation by developing a competency profile for S&T managers.

The working group was co-chaired by EC and the Department of National Defence. EC already had some recent experience in this area, having developed such profiles for various levels of management in its National Water Research Institute.[58] By December 1998, the working group had completed its task. It produced a report containing a competency profile as well as a list of learning opportunities for each competency and a number of recommendations about how best to move forward.

Developing science managers continued to be one of the main areas of interest for the ADMs' Steering Committee. Its agenda was limited by the different approaches departments took to management development. A few departments – Health Canada and the National Research Council, in particular – had their own management development programs. The majority, including EC, merely urged staff to undertake self-directed management development, with few if any requirements laid out or resources made available to support their learning activities. In addition, the Steering Committee did not see itself as being in the business of

---

[56] R. G. Lawford & G. Roy, *Results of a Needs Identification Survey for a Course for DOE Research Managers*, Corporate Planning Group, February 21 1980
[57] *Report of the Management and Scientific Development and Training team*, September 27 1996. EC Science Policy Branch, box 40
[58] *Framework for Competency Development*, 1998. Associated with this, NWRI also prepared a *Competency Dictionary* and a *Development Resource Guide*.

offering training. The most attractive option was to develop a relationship with the Canadian Centre for Management Development, the federal government's executive development body.[59] In collaboration with the Steering Committee, the Centre made use of the S&T manager competency profile. It tailored its web-based tools to include a self-assessment questionnaire so that S&T managers could evaluate their current competencies. The Centre also developed a map linking those competencies to learning options it offered.

Most significantly, the Centre worked with the S&T Community to develop a new course, *Leading Scientific Teams*.[60] Following up on the competency profile, the scientific management development working group had identified a common need among its member departments for a course for first-time supervisors of S&T work. Design work began in 2001, led by the Centre but involving departmental advisors. A pilot was offered the following February. It proved to be successful. Assisted by a guarantee from the Steering Committee for a minimum registration, the Centre began to offer the course on a regular basis.

Another opportunity to work on developing S&T managers presented itself in 2006. The Leadership Network, at the Public Service Human Resources Management Agency, formed a partnership with the S&T Community to fund a leadership development pilot program for scientists. The program took a small number of scientists nominated by their departments and engaged them in a series of classroom and experiential exercises to develop their management competencies. EC had three individuals in the pilot's first cohort and four in the second and last cohort. Although the program was judged a success, it could not be sustained by the departments. They found its costs too high. An ongoing recognition of the desirability of increasing the number of members from the S&T community moving into executive appointments in the federal government has remained, but there has been no further interdepartmental work on the issue.

Recruitment and Employment Equity

With a few exceptions – such as trained meteorologists – EC usually had little difficulty in recruiting S&T staff. It was an issue for the departmental management committee only once before Program Review in the mid-1990s. The NRC had decided in 1975 to discontinue its program of postdoctoral fellowships in favour of a system of research

[59] Now known as the Canada School of Public Service
[60] The Centre had earlier offered a course, *Managing Scientific Organizations*. However, it was no longer available by the late 1990s.

associates. It gave notice to other science-based departments and agencies, which had been using the program, that it would also soon cease administering the fellowships for them. The Science Advisor asked, in August 1976, for the departmental management committee's support for an EC program, but was refused.[61] However, the Department did step forward six months later, with the Office of the Science Advisor managing the interdepartmental program on a cost-recovery basis.[62] At the time, it consisted of 90 fellows in 10 departments and agencies.[63] This situation only lasted for one year. In April 1978, the program was transferred back to the NRC and soon became the responsibility of the newly created Natural Sciences and Engineering Research Council.

Program Review saw the dismissal of a large number of federal S&T employees. This gave rise to the question of whether the government would be able to attract high quality scientists, engineers, technicians and technologists into its workforce in the future. Not surprisingly, recruitment was one of the central concerns of the 1995 Framework for S&T human resources. Closely associated was the issue of the representativeness of federal employees in comparison with the composition of the Canadian labour force. In 1997 the Framework's Steering Committee struck a special working group on recruitment, retention and diversity. Its first task was to analyze the demographics of the federal S&T workforce. The result foresaw no crisis over the next five years in the numbers of federal S&T employees, although it did raise some concerns about the increasing number of departures projected due to the retirement of the baby boom generation.[64] This was to be the first in a series of demographic analyses sponsored by the Steering Committee, each of which would come to essentially the same conclusion.

The working group then turned its attention to the four employment equity groups: women, visible minorities, the disabled and aboriginals. Its findings showed that while representativeness was an issue in the federal S&T workforce, the nature of the challenge varied for each equity group and the recruitment needs were different for each department. Coming up with an initiative that could be sponsored through a community-wide effort would be challenging.

---

[61] Meeting, August 5 1976. LAC Acc. 1991-92/017, box 17

[62] Memo, February 18 1977. LAC Acc. 1993-94/003, box 4

[63] Memo, February 24 1977. LAC Acc. 1993-94/003, box 20

[64] *Estimates of Hiring Potential – Scientific and Technical (S&T) Community, 1998-2002*, prepared by Kathryn McMullen and Hara Associates, December 1997. EC Science Policy Branch, box 43

A good example of the need for targeted approaches to the issues facing equity groups is provided by the work done on recruiting aboriginals into the federal S&T workforce. The working group had determined that the biggest challenge in recruitment was the small supply of aboriginals trained in S&T.[65] EC worked with Natural Resources Canada to develop a funding program to encourage aboriginals to study S&T and to promote awareness of the federal government as an employer of choice within that community. The result was a three-year memorandum of understanding among seven departments providing $120,000 per year. The project started up in 2001, and was managed by EC and delivered initially through two aboriginal science associations.[66] The agreement proved to be successful and was renewed for a second term. It was discontinued after 2007 due to complications arising from a government freeze on grants and contributions, but also due to a belief that the agreement was no longer useful. By that time aboriginals were no longer underrepresented in the federal S&T workforce.[67] The ADMs' Integration Board decided in 2009 to no longer have a separate initiative devoted to aboriginals but to deal with all employment equity group issues through its regular task forces.

The issues facing women in federal S&T also attracted considerable policy attention. The Steering Committee set up a special working group on this subject in 1998, chaired by Karen Brown of EC. The group conducted a demographic analysis of women in the federal S&T workforce, an exit survey to determine reasons for departure, an analysis of university graduates, and a review of federal human resources policies around systemic discrimination towards women.[68] Parallel to this effort, EC also established its own working group, *Women, Environment, Science and Technology*, co-chaired by Karen Brown and Nancy Cutler. Financed through the Department's Learning Fund, this work was meant to complement that being done interdepartmentally by focusing on EC's situation. The working group distributed a questionnaire within the Department and followed it up with consultations with employees across

---

[65] *Report, Aboriginal Sub-Group of the Recruitment, Retention and Diversity Working Group*, May 1998. EC Science Policy Branch, box 43

[66] The Canadian Aboriginal Science and Technology Society (CASTS) and the Canadian Aboriginal Science and Engineering Association (CASEA). After the first year, the agreement was only with CASTS.

[67] *Aboriginal Employment within the Federal S&T Community*, discussion paper for the Aboriginal Youth Initiative (AYI) Working Group, prepared by Science Policy, EC, October 2008

[68] *Draft Report*, Women in Federal Science and Technology Working Group, March 1999. EC Science Policy Branch, boxes 47 & 48

the country through 15 roundtables. It made a dozen recommendations, which were accepted by the Departmental management committee.[69]

There was another revival of activity around women in federal S&T by the ADMs Committee in 2003. It sponsored preparation of a chapter on the public service in a handbook for women in science and engineering.[70] This led to the development of a one-day course dealing with the challenges women in S&T face in becoming leaders within the federal government. The course proved to be very popular and has been offered on a regular basis since 2007 by the Canada School of Public Service.

The Steering Committee also continued to explore various general recruitment initiatives. The work on the Graduate Opportunities Strategy in 2000-01 has already been noted above. The Committee also experimented with the establishment of government-wide inventories or pools of candidates. A good example is the 2003-04 pilot for entry-level positions, which the S&T Community sponsored with the help of the Public Service Commission. The intention was to shorten the length of time needed to staff positions by creating a pre-qualified pool of individuals from which departments could select. It was thought that the creation of such an inventory would also help to promote government S&T careers to university graduates. The pilot and subsequent versions of it were not very successful. The main challenge was in getting departments to make use of a general inventory. The experience highlighted once again the difficulties of trying to take collective, interdepartmental action in areas that are ultimately the responsibility of individual departments.

---

[69] *Report*, Women, Environment, Science and Technology (WEST), June 1999. EC Science Policy Branch, box 46

[70] F. Mary Williams & Carolyn J. Emerson, *Becoming Leaders: A Practical Handbook for Women in Engineering, Science, and Technology* (American Society of Civil Engineers, 2008)

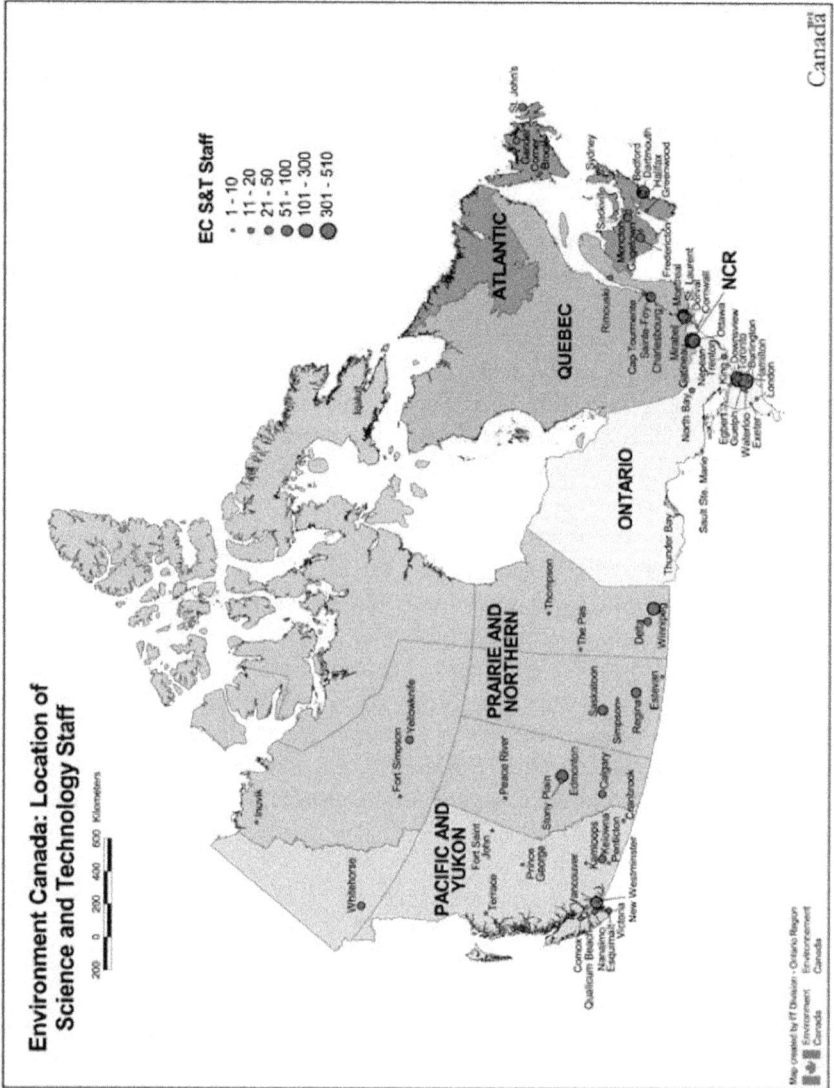

Figure 6.5 Geographic distribution of EC S&T staff, 2004[71]

---

[71] *Science and Technology: The Foundation for Policy, Regulation and Service*, EC, 2004

PART FOUR

COLLABORATION

7

Fostering Collaboration with Federal Departments and Agencies

At the time of its creation, EC was envisioned as playing a coordinating role in the federal government with respect to environmental issues. Many federal environmental activities, especially those related to specific natural resources, were to be found in other departments due to their mandates. The responsibilities of EC's Minister included "coordination of the policies and programs of the Government of Canada respecting the preservation and enhancement of the quality of the natural environment".[1] However, in practice this authority was seldom used. Interdepartmental relations tended to be quite competitive, so other departments were usually engaged in a collegial rather than dictating way.

Despite the challenges, EC's policy efforts for S&T frequently struggled with enhancing interdepartmental collaboration. This interest stemmed from the crosscutting nature of environmental issues and from the increasingly collaborative nature of the scientific enterprise. In addition, tight financial resources for S&T in EC as well as in other departments put a premium on leveraging and finding synergies in attempting to respond to a growing number of environmental priorities. This was especially evident in the decade after Program Review when there were several efforts, many of them led by EC, to foster greater collaboration among federal science-performing departments.

The attempts to increase collaboration revealed its inherent challenges.[2] Various administrative practices, the lack of incentives for collaboration and the organization of the federal government all served as barriers. EC's policy efforts were usually carried out in interdepartmental fora, and sometimes were focused on the full range of federal science, not just on environmental S&T. The experience tells a great deal about what could and more often could not be done in the absence of changes to the governance machinery of the federal government.

---

[1] *Powers, duties and functions of the Minister*, Department of the Environment Act

[2] *Moving from the Heroic to the Everyday: Lessons learned from leading horizontal projects*, Canadian Centre for Management Development, Roundtable on the Management of Horizontal Initiatives, 2001

Early Attempts to Coordinate Environmental Research

One week after the Speech from the Throne that announced the government's intention to create EC, a new National Research Council (NRC) associate committee met for the first time.[3]   The Associate Committee on Scientific Criteria for Environmental Quality had been proposed by the NRC and agreed to by the government in response to resolutions by the provinces to coordinate efforts to fight pollution and to establish minimum standards for pollution control.[4] Its purpose was to quantitatively assess, evaluate and disseminate information on the cause and effect relations of pollutants on receptors. It had seven subcommittees – air, water, metals, pesticides, physical energy phenomena, biology, and one devoted to trouble shooting – and involved scientists from across Canada.[5]

The establishment of the Associate Committee so close to the founding of EC highlights the fact that the government did not see the new Department as having responsibility for all environmental activities. As Prime Minister Trudeau said the day after the Throne Speech:

> The fight against the pollution of our environment is far beyond the capacity of one minister and his department. Indeed, it cannot be waged effectively by the federal government alone, or the provinces individually, or even just by Canada. It is a fight that must be waged by all ministers, all governments and all people.
>
> There are many departments in the government which have and will continue to have important responsibilities for the preservation of the quality of our environment. These departments will cooperate with the proposed Department of the Environment which will have the principal tools to lead the fight against pollution and to help co-ordinate the efforts of others.[6]

Departments such as Agriculture, Health, Transport, Indian and Northern Affairs, Energy Mines and Resources, Industry, and the NRC were all expected to maintain their environmental programs. EC's role

---

[3] Associate Committees were one of the main ways in which the NRC tried to coordinate science performed by universities, the government and industry. D. J. C. Phillipson, *Associate Committees of the National Research Council of Canada, 1917-1975* (National Research Council of Canada, 1983)
[4] G. C. Butler & H. H. Harvey, *A Twenty-year History of the National Research Council Associate Committee on Scientific Criteria for Environmental Quality (1970-1990)*, NRC 1990
[5] Ibid.
[6] House of Commons, Debates, October 9 1970, p. 35

was to oversee environmental quality, coordinate federal programs and cooperate with the provinces and other bodies.

In order to help exercise its coordinating mandate, EC decided in May 1972 to ask Cabinet to establish an Interdepartmental Committee on the Environment, which its DM would chair.[7] The Committee held its first meeting about a year later. In the fall of 1974, the NRC proposed that a subcommittee on research be created.[8] The Interdepartmental Committee formed such a group in June 1975, with EC's Science Advisor as chair. Its purpose was to be a forum for the coordination of environment-related scientific activities.[9] The Sub-Committee on Research held its first meeting on November 14th.[10] At its second meeting, the following March, information on selected projects was shared, covering 29 initiatives from 14 departments.[11] However, there was not much interest in the subcommittee and it soon became dormant.

The NRC's Associate Committee had also touched on the issue of coordination. It had identified research priorities and exchanged information on research done outside the federal government.[12] At its peak, it had about 100 scientists involved in its work.[13] However by the mid-1980s it found itself on the periphery of the NRC's interests, which under tight budgets were being increasingly restricted to assisting Canadian industry. Despite a letter from EC's DM to the NRC's president noting the importance of NRC's environmental work, the Associate Committee was effectively shut down in November 1984.[14] Its funding was slashed from $1.1 million to $15,000 for 1985-86, and its secretariat at the NRC closed.[15] The Associate Committee continued to exist until 1990 when it was finally abandoned, but it undertook no new work, finishing what was already underway. EC provided coordination assistance to the Committee on a part-time basis.

The fate of the Associate Committee and of the Sub-Committee on Research reveals some of the challenges facing coordination in the federal government. The tendency of departments to operate in isolation of one another, amplified by the weaknesses of government institutions for

---

[7] Meeting, May 11 1972. LAC Acc. 1991-92/017, box 3

[8] Letter from EC DM, November 13 1974. LAC Acc. 1993-94/003, box 15

[9] *DOE Submission to the Senate Committee on Science Policy*, January 13 1976. LAC Acc. 1993-94/003, box 27

[10] LAC Acc. 1993-94/003, box 2

[11] LAC Acc. 1993-94/003, box 15

[12] Memo, November 14 1975. LAC Acc. 1993-94/003, box 2

[13] Op. cit., footnote 4

[14] Briefing note, December 18 1985. EC OSA, box 1

[15] Its budget increased to $35,000 in 1986-87, and was $30,000 annually after that. Ibid.

coordination, was worsened by a focus on core missions brought about by financial restraint.

Board on Environmental Research and Technology

The deteriorating financial position of the federal government in the 1980s led to a number of studies on where savings might be found, and these in turn generated calls for increased coordination. For example, the new Progressive Conservative government in 1984 appointed a Task Force on Program Review, led by Erik Nielsen, aimed at reducing government expenditures. Its report on EC's programs recommended that the Department look at the interdepartmental Panel on Energy R&D as a model for coordinating, setting priorities and managing environmental science.[16] The report was soon followed by a second study, the *Environmental Quality Strategic Review*.[17] It called for improvements to the management and coordination of federal environmental S&T, and proposed a pilot coordination mechanism to be run through MOSST, with EC setting environmental S&T priorities. Despite some attention by the media, these reviews did not have much impact on the coordination of environmental S&T.

However, the idea that relevant S&T in various federal departments and agencies ought to be better linked and mobilized for the environment persisted. It appeared again in planning for the government's new environmental agenda, the Green Plan, in early 1989.[18] At that time, a Board on Environmental Research and Technology was proposed.[19] As recommended by the Nielsen Task Force, it was modelled on the Panel on Energy R&D.

> The pervasive nature of environmental S&T is similar to that of energy R&D, where the federal government possesses, in the Panel on Energy R&D (PERD), a proven and highly effective mechanism for promoting and coordinating multidisciplinary and multi-agency science, for redirecting

[16] *Programs of the Minister of the Environment*, 1986. The review team was led by George Layt, senior vice-president at Stelco. The Panel on Energy R&D had been in operation since 1975.

[17] Led by Alain Desfossés, a special advisor in the Privy Council Office, and published in 1986.

[18] Note from John Hollins to Peter Fisher re framework for Environmental Action Plan, March 3 1989. EC OSA, box 1

[19] I am using the name that was finally settled on, in order to avoid confusion. Very early proposals sometimes used Board on Environmental S&T or Interdepartmental Panel on Environmental S&T.

resources on an annual basis to meet new priorities, and for forging partnerships with other governments, universities, and the private sector.[20]

The role of the Board would be to develop a strategy for environmental S&T, provide funds for priority issues, and manage federal environmental science in an integrated way. Like the Panel, it would be organized under an ADM-level committee and consist of a number of tasks and committees (see Figure 7.1). Environment Canada would chair the Board and provide its secretariat. The principal departments involved would be the large science-based departments: Energy, Mines and Resources, Fisheries and Oceans, Health and Welfare, Agriculture, Canadian Forestry Service, and the NRC. Initial thoughts about the Board's budget placed it at $607 million over 5 years. By July of 1990, this had been scaled back to $180 million – which was still the largest of the proposed six elements of the S&T component of the Green Plan.

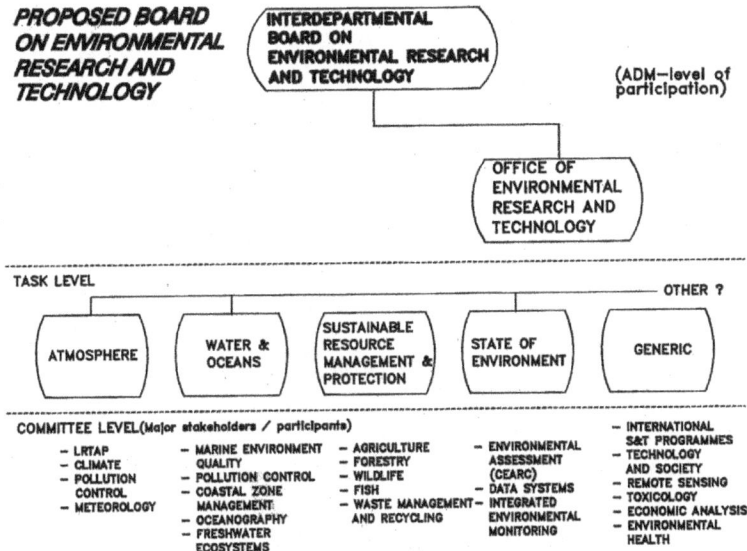

Figure 7.1 Proposed organization of the Board of Environmental Research & Technology

The Board did not become an initiative under the Green Plan. Despite support during the consultations on the Plan and from the

---

[20] *Environmental Science Partnerships for the Future*, July 1989. EC OSA, box 1

Office of the Science Advisor, in mid-October the Office was informed that the Board would not be included in the Green Plan.[21] This occurred just days before a scheduled workshop to provide advice on how the Board should function.

The reasons for not proceeding with the Board are unknown. However, it does appear that other departments were not that keen on the proposal. The Office of the Science Advisor decided to go ahead with its planned workshop, but to use it to see if there was a consensus about the need for coordination of environmental S&T. Participants could not reach such an agreement. Departments were reluctant to have Environment Canada do anything more than gather information.

> They were uncomfortable with the word "coordination" which seemed to smack of management and interference. They preferred the concepts of reporting and cooperation among the players, using whatever mechanisms were available, and were prepared to leave the control of research decentralised.[22]

Developing an integrated management plan for environmental S&T in the federal government would need to address the barriers to collaboration within the federal system, most particularly the inclination of departments to control their own turf.

MOU on S&T for Sustainable Development

Just as the Green Plan had provided an opportunity to propose a mechanism to support collaboration in federal environmental S&T, so too did the election of a new Liberal government in October of 1993. During the election campaign, the Liberal Party released its policy platform – known as the Red Book – which prominently featured the concept of sustainable development.[23] The Environment Canada Science Committee had been considering ideas for several months about a second

---

[21] *A Framework for Discussion on the Environment*, 1990. See also *Green Plan Consultation. National Wrap-Up Session. Workshop Reports*, August 1990. EC OSA, box 2

[22] *Draft Report on the Workshop on Environmental Science and Technology sponsored by the Science Advisor's Office of Environment Canada. October 29th and 30th 1990.* See also *The Way Ahead: A suggested approach to further action on environmental science and technology following the workshop at EcoNiche 29th-30th October 1990*, by Alan Cairnie and Max McConnell. EC OSA, boxes 2 & 5

[23] Chapter 4 in *Creating Opportunity: The Liberal Plan for Canada*. This was elaborated in a follow-up backgrounder, *Sustainable Development: Environmental Challenge; Economic Opportunity*.

departmental science forum, but none had generated any great enthusiasm.[24] With a new government in place, the Committee quickly decided that the forum would focus on science for sustainable development.[25] Not only was this topic central to the role of EC's S&T, it was also one that would engage other departments in thinking about linking their S&T activities in order to deliver on the government's agenda.

EC's *Science Forum '94* took place on May 9-11, 1994. By design, half of the 90 participants came from other departments. Its objectives were to develop a common understanding of the government's agenda on sustainable development and its implications for S&T; to discuss the elements of an interdepartmental S&T strategy to support that agenda; and, to identify the contributions that departments could make to that strategy.[26] One of the meeting's main conclusions was the need for departments to develop a government-wide strategy for S&T for sustainable development and to "create a forum which would identify priorities on an ongoing basis and propose new mechanisms for managing S&T in a more integrated fashion," perhaps modeled on the Panel for Energy R&D.[27] Participants also saw the recent government announcement of the Federal S&T Review (in February) as an opportunity to further promote and refine their suggestions.

The Forum provided the basis for a rather swift response to a challenge from the Clerk (Jocelyne Bourgon). At a meeting in the early fall of 1994, noting that the DMs of the natural resource departments were all talking about conducting S&T for sustainable development, she asked how that effort was coordinated.[28] EC's DM, Mel Cappe, started up a DM committee to look into strengthening coordination. The group's first meeting reviewed a paper, *Framework for Sustainable Development S&T in the Natural Resource Sectors*. Robert Slater, ADM of the Environmental Conservation Service and EC's lead for science policy, then called a meeting of his counterparts in Agriculture, Fisheries & Oceans and Natural Resources, over lunch on December 20th. They decided to develop a memorandum of understanding to implement the Framework. By January 18th, the DMs of the four departments had signed it.

The *Memorandum of Understanding between the Four Natural Resource Departments on Science and Technology for Sustainable Development* aimed at

---

[24] The first Science Forum was held November 30 – December 2 1992 and focused on providing scientists' views on strategic directions for R&D. EC OSA, box 3

[25] Minutes, Environment Canada Science Committee, January 12 1994. EC S&T Management Committee, box 1

[26] *Report on Science Forum '94*. EC OSA, boxes 3 & 6

[27] *Science Forum '94. Synthesis Report*

[28] Briefing note, John Hollins. EC MOU-S&T, box 1

optimizing the use of departmental S&T resources to achieve sustainable development goals. The 3-year agreement set out how the departments would manage collaboration, through ADM and DG committees. It also identified five issues for joint action: priority setting methods, coastal zone management, renewable energy, climate change, and metals in the environment. Almost immediately, a sixth issue was added: ecosystem effects of ultraviolet radiation. Even though the MOU did not require it, the first set of issues was selected so that all four departments participated in each of the six working groups. EC was strongly involved, leading three of them.[29] Over the following years, many other working groups would be added. Overviews of their efforts can be found in the annual reports of the Memorandum of Understanding.[30]

The Memorandum of Understanding was renewed – essentially unchanged except for the addition of a fifth department, Health Canada – for another 3-year period in 1998.[31] Although there was no formal renewal in 2001, its work continued for about two more years until early 2003 when the DGs' committee decided to recommend that the MOU not be continued.

Over its 8-year life span, the Memorandum of Understanding fostered a number of achievements, mostly found in the results of its scientific working groups. However, from the perspective of a mechanism to increase collaboration, it had limited success. While it provided a structure for departments to meet regularly and to take collective action on specific topics, the Memorandum hardly ever went beyond coordination of existing activities. The 1998 Auditor General's report critiqued it for not including joint goal setting and planning.[32] The ADMs attempted to go further, but were unable to develop even a common business plan.

Established at the time of Program Review, the Memorandum of Understanding always suffered from a lack of financial resources, which often limited the scope and scale of its projects. Reliant on existing resources, its projects competed with departmental priorities for the time of staff. The lack of resources was also not conducive to moving beyond the project-based model of the initial Memorandum to a more strategic planned approach. Moreover, it became increasingly clear during its later years that the emphasis on scientific collaboration often meant that the

---

[29] Metals in the environment, climate change, and ecosystem effects of ultraviolet radiation.

[30] Annual reports were issued for the first four years, followed by two biennial reports.

[31] The MOU itself was almost identical to the 1995 version. However, the Framework was rewritten to reflect some of the lessons that had been learned in the first three years.

[32] *The Federal Science and Technology Strategy: A Review of Progress*, chapter 22 in the Report of the Auditor General of Canada to the House of Commons. December 1998, p. 13

links to policy were neglected. This situation led to problems moving the issue-based science effort forward due to the absence of a policy pull for those issues. What was needed, but missing, was attention to increasing collaboration on the policy side.[33] Ironically, the Memorandum of Understanding came to an end at a time of renewed interest and a new vision for interdepartmental S&T collaboration, once again led by EC.

FINE

In May of 2001, a group of ADMs from the largest science-performing departments met to discuss the positioning of federal S&T in the unfolding innovation strategy. The latter was a follow up to the January Speech from the Throne which had announced that the government would build a world-leading economy driven by innovation, ideas and talent and had set a target of moving Canada from 15th to 5th place in world R&D. A draft white paper had been prepared by Industry Canada and by Human Resources Development Canada, which included only a brief reference to strengthening the research capacity of federal departments and this in the context of funds for research facilities. EC's DM, Alan Nymark, asked his staff to prepare some alternative text that would better address the various challenges facing federally performed S&T. This text became the basis for discussion by the group of ADMs, led by Robert Slater, then EC's Senior Assistant Deputy Minister.

The ADMs quickly agreed on an outline for a federal "Knowledge Investment Strategy" whose goal was to leverage existing S&T capacity in federal departments in order to mobilize it toward national priorities requiring collaboration. The Strategy would have four components: capital investment in federal S&T infrastructure, recruitment of federal S&T talent, cooperative S&T networking for public policy, and a targeted federal network on wealth creation.[34] The ADMs met almost weekly over the summer to guide the development of the Strategy and to flesh out its four components.

The ADMs had for some time been concerned with the issue of S&T capacity. They had all dealt with substantial cuts to their S&T efforts during Program Review. They had been engaged in discussions over the

[33] For a fuller discussion of the lessons learned about the Memorandum of Understanding see the report commissioned by Environment Canada from the Impact Group, *Role of a Renewed 5NR MOU in the Evolving Spectrum of Horizontal Federal S&T Management*. Science Policy Branch working paper #18, December 2002
[34] *Knowledge Investment Strategy: Strengthening the S&T Capacity of the Federal Government in Support of Innovation*. Science Policy Division files

last couple of years on the issue, which had not led to much success. Despite increasing amounts for S&T in recent federal budgets, science-based departments had been unable to obtain much new funding for their priorities. Almost all of it was going to universities.[35] The ADMs had also been involved in the work of the Council of Science and Technology Advisors, an external advisory group to the government on federal S&T. One of its recent reports had called for the government to implement and fund new models for S&T that moved away from a departmental approach to a more horizontal (across departments and the innovation system), competitive, multi-stakeholder approach.[36] This recommendation reflected the views of several S&T ADMs at that time. They believed that they needed to come up with innovative proposals for investments in federal S&T in order to capture the interest of central agencies and of the government.

A proposal for a "Knowledge Investment Strategy" was presented at a working dinner meeting on August 23rd of the DMs of the eight departments that had been involved in developing it.[37] The DMs liked the Strategy but felt that its four components made it too complicated. They also thought that it would have a greater impact if it included a short list of compelling examples of important policy issues that could be addressed through the Strategy. By mid-September the Strategy had become the Federal Innovation Networks of Excellence (FINE), and an interdepartmental workshop had been organized to identify issues.[38]

FINE was modelled on the Networks of Centres of Excellence program and on the Toxic Substances Research Initiative.[39] Under FINE – in contrast to the Networks program, but similar to the Initiative – federal departments would lead the networks and could obtain funding for their participation. Like the Strategy proposal, FINE's goal was to integrate science capacity inside and outside the government and to bring it to bear on emerging policy issues and opportunities for economic

---

[35] New funding for university R&D announced in budgets 2000 and 2001 totalled over $8 billion (spread out over a number of years).

[36] *Building Excellence in Science and Technology (BEST): The Federal Roles in Performing Science and Technology,* 1999, p. 28

[37] EC, Agriculture & Agri-Food, Fisheries & Oceans, Natural Resources, Health, Industry, National Defence and NRC

[38] The workshop was held on September 20th and 21st. It identified the following seven issues: stewardship in sensitive environments, emerging threats and opportunities in the Canadian food supply, water quality and aquatic resource protection, assessing the products of biotechnology, protecting Canadians from emerging threats to national security, information systems security and access for all Canadians, and tools and infrastructure to deliver FINE.

[39] The Networks of Centres of Excellence program was launched in 1989. The Toxic Substances Research Initiative was created in 1998 and operated until 2003.

development. It provided an alternative way of managing federal S&T, one that was cross departmental, competitive, expert reviewed, funded for fixed periods of time, open to non-governmental participants, and focused on emerging national priorities.

A good example of the potential of FINE was the creation of a program that would later be named the CRTI (Chemical, Biological, Radiological-Nuclear, and Explosives Research and Technology Initiative). The impetus for it came from the September 11 2001 terrorist attacks and destruction of the World Trade Center in New York. However, its design was based on the ideas behind FINE. Indeed, an early presentation on that initiative was titled "FINE for National Security and Counter-Terrorism."[40] The CRTI's champion, John Leggat (the head of Defense R&D Canada), had been an active member of the ADMs Committee on FINE. The CRTI received $170 million over five years in the December federal budget.

Despite the CRTI's success, another year was spent in developing the details of FINE. In particular, much effort and thought were given to its governance system and to engaging the interest and support of potential participants in FINE.[41] For example, the ADMs Committee spearheaded the Federal S&T Forum in October 2002. Participants there heard from a panel of DMs on the future of federal S&T, were asked to help revise the vision for federally performed S&T which had been drafted under FINE, and brainstormed about actions to improve horizontal management of federal S&T.

By the end of 2002, a much laboured over memorandum to cabinet on FINE had been drafted. However, the proposal went no further. At least one DM, Samy Watson (then at Agriculture & Agri-Food), now felt that FINE did not go far enough. He believed that if approved it would stand in the way of more fundamental change to the integration of S&T across departments and with policy development.[42] In addition, Slater found it very difficult at that time to obtain Ministerial support.

Integration Board

Although FINE had come to a standstill, its ADM steering committee continued to meet. Co-chaired by John Leggat and Karen Brown, EC's

---

[40] Presentation to DMs meeting, October 15 2001. Science Policy Division files

[41] See, for example, the presentation *FINE: Governance and Management Issues*, by Jim Mitchell, September 23 2002. Science Policy Division files

[42] Expressed in his comments as a member of the DM Panel at the 2002 Federal S&T Forum.

ADM of the Environmental Conservation Service, the committee was mostly concerned in early 2003 with following up on the Federal S&T Forum. However, prompted by Leggat at its April 16th meeting, the committee decided to take a bold step forward. Noting that many other countries had government S&T coordinating mechanisms, they decided that governance was central to their ability to foster improved S&T collaboration among federal departments.[43] If no new funding could be obtained for crosscutting S&T initiatives, then the ADMs would explore a governance structure for coordinating existing S&T capacity that did not rely on additional resources. The group decided to rename itself the S&T Governance Committee and to organize a workshop to further refine its role.

The workshop, "Integrating S&T across Science Based Departments and Agencies," was held on June 10th and 11th in Merrickville, a village just outside Ottawa. About 45 people from 10 departments participated, including ADMs, senior science managers and policy executives. The meeting had two results. The Committee gave itself a mission to "provide strategic leadership, guidance and direction for mobilizing and integrating S&T efforts across departments and disciplines, focusing on the priorities of Canadians."[44] The Committee also once again renamed itself, now as the S&T Integration Board, to better reflect that mission. The Board had 10 member departments – EC, Fisheries & Oceans, Natural Resources, Agriculture, Health, Industry, NRC, Transport, National Defence and the Food Inspection Agency – with each contributing about $25,000 to support its activities as well as a one-person secretariat.[45]

The second output of the workshop was a decision to undertake a process to identify areas that would benefit from joint S&T activities by departments. S&T collaborations among departments were not unknown at that time. An inventory of them, prepared for the Board, revealed 105 such collaborations adding up to $458 million or 11% of total federal intramural S&T spending.[46] EC was one of the largest spenders on these collaborations and one of the top departmental collaborators. It was clear that many important issues facing the government did not fit neatly within departmental boundaries. What was new in the formation of the Board was a coordinated effort to horizontally manage the whole set of

---

[43] The EC commissioned paper, *International Comparative Study of Approaches Used to Address Issues that Cut Across Science-Based Departments*, was the basis for their discussion. Science Policy Branch working paper #23, March 2003

[44] *A Strategic Plan for the ADM S&T Integration Board*, Science Policy Division files

[45] William Doubleday, who had been brought to EC on assignment from DFO to help with FINE, took on the role of secretariat for the S&T Integration Board.

[46] Presentation on *Federal Collaborative S&T Initiatives as of Fiscal Year 2002/2003*, June 1 2004. Science Policy Division files

issues, and to do so without the incentive of new funding. The idea was that the Board would provide leadership and serve as a catalyst but not actually manage existing or new joint initiatives. That would be left up to the lead departments.

A strategic planning workshop was held on November 28th, attended by over 50 science and policy managers. It was organized around issues that had been identified as potential candidates for integration since the Merrickville meeting, and was looking for departmental commitments to lead or participate in working groups to tackle those issues. The result was agreement on five topics for which integration plans would be developed. EC was the lead on three: water, wildlife diseases and invasive species. The other two were northern S&T (led by Fisheries & Oceans and Indian and Northern Affairs) and climate change (Natural Resources). Three months later, at a Board meeting on February 24th 2004, each of the five working groups reported on its progress.

Despite a strong start, the Integration Board faced a number of challenges in its attempt to engage the full capacity of federal S&T, as noted by the Council of Science and Technology Advisors:

> Although the world has changed significantly, the bureaucratic structures that underpin the government's management of its internal S&T have remained relatively unchanged. The federal system is still dominated by traditional vertical departments which are structured largely to provide S&T-based solutions to issues within their specific jurisdictions. The accountability, resource allocation and reward systems characteristic of this type of vertical system lack the incentives, flexibility and responsiveness that facilitate horizontal S&T.[47]

One significant challenge for many of the S&T ADMs was the absence of higher-level S&T governance. The Integration Board had been an initiative of a group of ADMs from S&T performing departments. It did not have any formal upward reporting relationship. When the government re-created the position of National Science Advisor and appointed Arthur Carty to it in April 2004, the Board saw an opportunity to gain greater authority. Carty was well aware of the currents in federal policy for science, having been President of the NRC. He announced in June his intention to create a DMs committee to ensure coordination and complementarity of federal investments in S&T.[48] This was repeated in

---

[47] *Linkages in the National Knowledge System: Fostering a Linked Federal S&T Enterprise*. Report of the Council of Science and Technology Advisors, February 2005, p. 2

[48] June 28th meeting of the ADMs Committee on S&T. Science Policy Division files

the October Speech from the Throne.[49]   Between October and January, he organized three dinner meetings of the DMs of science-based departments.

Carty acted as a catalyst for higher-level support for increased S&T collaboration. For example, the Integration Board used his meetings to obtain the DMs' support of and engagement in the January 2005 Federal S&T Forum, "Moving from Collaboration to Integration." It attracted over 350 participants, the large majority of whom were enthusiastic about the benefits of integration. However, Carty also served to deflect the Board from its core mandate. For instance, he wanted the Board to develop a report on the mission critical science in each department, and he got it to work on a laboratory infrastructure proposal.[50]

The experience with the DMs also brought to mind, for those involved with the Integration Board, the old adage "be careful what you wish for as you may receive it." While generally supportive of the Board, the DMs' involvement could also paralyze it. For example, building on Carty's meetings, the Treasury Board Secretary called a meeting in April 2005 to discuss some shared management topics affecting science-based departments. From this meeting emerged two new DM-led initiatives, both involving the Integration Board. One was to look at the barriers to S&T collaboration, led by the President of the Food Inspection Agency.[51] The other was to identify emerging priorities for S&T integration, led by EC's DM, Samy Watson.

The EC task became a major workshop involving 44 federal scientists from 14 departments and agencies who met for 9 days between the 13th and the 23rd of September. They identified and explored eight national challenges requiring an integrated approach: climate and ecological knowledge; pandemic and infectious disease; dealing with preventable diseases; value from waste; sustainability and natural landscapes; water and food quality, quantity and security; clean air; and, alternative energy and the intelligent use of hydrocarbons.[52] The workshop participants also outlined a way forward for the management of S&T integration, which they later developed more fully.[53]   These reports were presented to the DMs' Policy Committee on Environment and Sustainability, which was

---

[49] The Speech from the Throne was delivered on October 5th.

[50] *Science and Technology in Support of Mission Critical Goals: The Work of the Science Based Departments and Agencies*, discussion paper, June 2005. Science Policy Division files

[51] The resulting report was *Overcoming Barriers to S&T Collaboration: Steps Towards Greater Integration*, March 9 2006. Science Policy Division files

[52] *Beyond the Horizon: Identifying Emerging Priorities for S&T Integration*, workshop report, September 13-23 2005. Science Policy Division files

[53] *Proposal for an Integrated Federal Science and Technology Response to National Challenges: A Blueprint for Action*, January 2006

chaired by Watson. It tasked the Integration Board with identifying a few initial areas to test the workshop's ideas.

The Board examined five areas in depth and selected freshwater research, public security, and pandemics / avian influenza. However, its report never made it back to the DMs. Their planned meetings on S&T were postponed after October 2005 and finally overtaken by events. In May 2006 the government announced, in its budget speech, plans for a new federal S&T strategy. Soon after, the National Science Advisor was moved to Industry Canada to report to the Minister of Industry.[54] S&T integration was firmly on the back burner, with the Integration Board in a wait and see position.

Apparently at a dead end, the Integration Board initiated in June 2006 a new project aimed at building the base for a government-wide approach to its own S&T activities. The Board decided to develop a map of those activities organized by the outcomes to which the S&T was targeted. The map was championed by Robert Walker, the CEO of Defence R&D Canada, who saw it as the first step toward collective management in those areas. It was developed through a series of six workshops in the fall of 2006 and winter of 2007. These determined that all of the government's internal S&T capacity could be categorized under five headings: economy, energy, environment, health, and security and defence (the environment category is shown in Figure 7.2).[55] The map included several levels of sub-categories in order to better describe the activities. The Board believed that the map would be a useful tool for identifying possible synergies or existing gaps in S&T activities devoted to similar outcomes but carried out in different departments. It would also be a useful communications tool about the purposes and scope of federal S&T.

In addition to the S&T Map, the Integration Board was involved in many other projects. Indeed, the original purpose of the Board – to enhance S&T collaboration – had become a minor activity. Meeting on average 18 times a year, the Board proved to be a convenient venue for ADMs of science-performing departments to share information and discuss areas of common management interest. The broadening of the Board's focus was reinforced in 2007 with the inclusion of the work on human resources issues of the former S&T ADMs Advisory Committee. The experience of the Integration Board revealed that while it was good at identifying issues, which might benefit from an integrated approach,

---

[54] The strategy was released in 2007 as *Mobilizing Science and Technology to Canada's Advantage*. The position of National Science Advisor was abolished in 2008.
[55] *Federal S&T Map of Outcomes and Activities: The contribution of key federal S&T performers to national outcomes*, March 2009

and successful at providing a forum for community building among the
S&T-performing departments, it was not effective in its mission to
mobilize and integrate S&T efforts across departments.[56] On its own, the
Integration Board proved to be insufficient to deliver its intended
results.[57] More fundamental changes to the governance of federal S&T
appeared to be necessary if the ability of departments to collaborate on
S&T was to be improved.

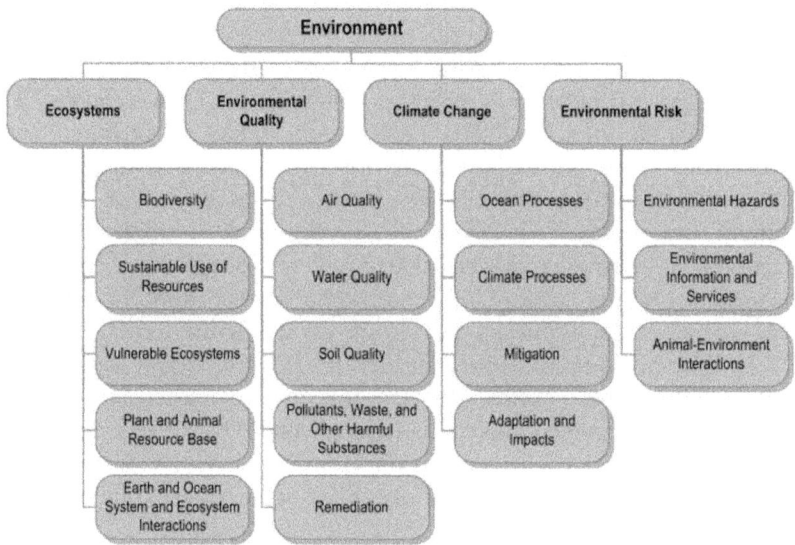

**Figure 7.2 Map of Outcomes & Activities of
Federal Environmental S&T, 2009**

---

[56] The S&T Integration Board continues to be active.

[57] The last report of the Council of Science and Technology Advisors concluded that a
number of changes to the existing structure and governance of federal S&T were required
if that S&T was to be better integrated. The report, *Facing Opportunities and Challenges
Underlying Society (FOCUS): Federal S&T Management in the 21st Century*, was never publicly
released before the Council was abolished in 2007. Some information on it can be found
in "Unreleased CSTA report calls for major changes in role and function of federal S&T,"
Re$earch Money vol 21, no 20 (December 21 2007): 3

8

Maintaining Links with Universities

Universities have always been important resources for EC. They educate many of its employees and they are important sources of expertise and of research for the Department. The relationship is not just one-way. EC supports university missions. It helps educate students through co-op and internship programs. Departmental employees sometimes give courses or guest lectures and supervise student projects. They hold collaborative positions in universities – about 200 in 2004.[1] There is also a close relationship between EC and universities in research. Almost 70% of EC's peer-reviewed scientific articles between 2003 and 2007 were co-authored with university faculty and students.[2] In addition, the Department supports university research through providing access to unique databases, large field programs and major facilities.

The close relationship usually emerged out of personal contacts and work in areas of mutual interest. But from time to time, S&T collaboration with universities moved beyond individuals and specific programs into the realm of Departmental policy. Since its formation, EC has sought to find ways to further develop environmental science capacity in universities, and to better align and link it to the Department's mission. EC has faced several challenges in this policy work. Insufficient funding, a decentralized department, a lack of understanding of the role of government science, inflexible governmental financial practices and the weak coordination of federal policy for science all presented barriers – sometimes too great to overcome – to the Department's plans. However, in its efforts to get more out of its scientific relationships with universities, EC proved to be very innovative and nimble if not entirely successful.

Science Subventions

Within a few months of EC's establishment, the departmental management committee decided to set up a task force to develop uniform guidelines and objectives for the Department's various grants

---

[1] *Smart Partners: Innovations in Environment Canada – University Research Relationships.* Science Policy Branch working paper #33, 2004
[2] See Figure 8.1.

programs.[3] At the time, EC spent $9 million on extramural research, about 11% of its total research expenditures.[4] Universities received $4 million, other public and private sector organizations the remainder.[5] The majority of the funding (70%) that went to universities was in the form of grants. It was administered through five grants programs – forestry, water resources, wildlife, fisheries and atmospheric – which had come into the Department when it was assembled from their respective organizations.

The task force extended work on evaluating the grants program of water resources, which had commenced earlier in 1971 before EC opened its doors. A consultants' report was prepared as was a discussion paper by the task force.[6] After several months of review, the Science Subvention Policy was approved in principle in August 1972 by the departmental management committee, and received Treasury Board approval in March 1973.[7] The Research Coordination Directorate was charged with general coordination of the Policy.

The Policy was based on the principle that research performed for the Department should be relevant to its mission. The objective of maintaining pools of expertise was to be left to the granting councils. In this respect the Policy reflected leading edge thinking at that time, which called for developing science policies and managing science in relation to economic and social objectives.[8] It replaced existing grants with research agreements to allow for greater control by the Department. It emphasized liaison between EC staff and recipients, and sought to support research that would be of mutual benefit.[9]

The EC Science Subvention Policy faced a number of challenges. It had difficulty in being a departmental program. The coverage of EC's interests was very uneven. For example, Water Resources spent $1 million through its program, Wildlife spent $25,000, and several Services

[3] Meeting, September 2 1971. LAC Acc. 1991-92/017, box 1
[4] H. F. Fletcher & A. E. P. Watson, *A Proposed Research Subvention Policy for the Department of the Environment*, Research Coordination Directorate [1972]
[5] The Water Management Service accounted for half of the $4 million, Lands Forests & Wildlife for $1 million, Atmospheric Environment for $0.5 million, and Fisheries for $0.4 million.
[6] The task force discussion paper is referenced in footnote 4. The consultants' report is T. J. Cartwright & M. Chevalier, *A Research Subvention Policy for the Department of the Environment – with particular reference to universities*, [March 1972].
[7] Meeting, August 23 1972. LAC Acc. 1991-92/017, box 4. See also LAC Acc. 1993-94/003, box 27.
[8] See for example, the Brooks Report, *Science, Growth and Society: A New Perspective*, OECD, October 1971. The report was circulated to EC's ADMs when it came out, and likely was used by the Task Force.
[9] LAC Acc. 1993-94/003, box 27

nothing. There was no way of dealing with environmental research that went beyond the responsibilities of individual Services since implementation of the Policy was left up to them. The Policy had aimed at funding social science and interdisciplinary research, but this did not happen. And one part of the Department bypassed the Policy. Forestry discontinued its subventions and put the money into block grants to the six Canadian forestry schools. The other major difficulty was financing. The annual amount spent on subventions throughout the 1970s was constant at about $2 million. EC's requests to Treasury Board for budget increases under its grants and contributions vote were turned down. The program could not expand into new areas.[10]

Other federal departments shared EC's desire to forge closer links with universities and encountered the challenges involved in doing that. In 1980, MOSST consulted departments about their ability to draw on university research. They reported a lack of sufficient funding and flexibility, as well as the need for better coordination with the granting councils.[11]

Innovative Partnering

EC set up a working group late in 1986 to once again examine the Department's relationship with universities. Out of this emerged a proposal for a departmental university strategy. Its objectives were to ensure a broad knowledge base to support EC's needs, the existence of centres of excellence outside government in priority environmental S&T areas, and the provision of trained graduates in the environmental sciences. The strategy acknowledged the variety of ways in which EC interacted with universities, but its emphasis for future directions was to work cooperatively with NSERC. In particular, it called for joint efforts in overcoming impediments to interdisciplinarity and international cooperation, and in forging links with the granting councils for the social sciences and humanities and for the health sciences. These new directions reflected the Department's experience with the knowledge needs of environmental issues as well as what some scholars were suggesting was

---

[10] This paragraph is based on two evaluations of the Policy done in the 1970s. *The Subventions Program: An Evaluation and Recommendations for Action*, Office of the Science Advisor, December 13 1974. *Departmental Subventions Program Policy – Evaluation*, March 1977. LAC Acc. 1993-94/003, boxes 29 and 4, respectively

[11] *Science Departments' Position Paper on University-Related Activities*, September 19 1980. EC Roots, box 7

the new style of scientific research in the late twentieth century.[12] In the fall of 1988, the Environment Canada Science Committee held a series of regional workshops to strengthen the Departmental network of staff concerned with the management of S&T. Included among the presentations was one on the university strategy. It was very well received, according to a report on the workshops.[13]

When an opportunity presented itself in 1990, in the form of the Green Plan, EC developed a very innovative program to follow up on its university strategy. The six-year, $50 million Eco-Research Program was announced in September 1991. It had three components: research grants for in-depth studies of Canadian ecosystems, research chairs and doctoral fellowships. The objectives of the Program were very similar to those in the strategy. Eco-Research sought to strengthen Canadian university research and training on environmental issues through supporting cross-disciplinary research, contributing fundamental and practical knowledge about environmental issues, making the research results widely available to Canadians, training specialists in environmental fields and encouraging national and international partnerships.

During early discussions about the Program, the granting councils – and NSERC in particular – expected that they would each receive a share of the new funding. But the Office of the Science Advisor was firm in its view that this would "pigeon-hold the funds and prevent the development of joint programs."[14]   After much negotiation, it was decided that a Tri-Council Secretariat run out of SSHRC would administer the Program. EC sat on both the management and operations committees as an observer.

Like other Green Plan programs, Eco-Research soon began to suffer from budget cuts, finally being cancelled due to Program Review. Eco-Research did operate over the six years from 1991-92 to 1996-97 and spent a total of $27 million: $19 million for 10 research grants, $4 million for 5 chairs, and $4 million for 89 doctoral students. A review of the Program's outcomes revealed that it had enhanced cross-disciplinary research in Canada, addressed some significant environmental research issues, encouraged the formation of networks of researchers that

---

[12] On the changing situation of science at that time see, for example, Michael Gibbons et al., *The New Production of Knowledge: The Dynamics of Science and Research in Contemporary Societies* (Sage, 1994).

[13] *National Workshops on Science and Technology Directions for Environment Canada*, February 1989. EC OSA, box 1

[14] Briefing note, December 13 1990. EC OSA, box 10

persisted after the Program ended, and had a significant effect on the careers of individual researchers.[15]

Eco-Research was not the only university-oriented program funded by the Green Plan. Two others were the Canadian Cooperative Wildlife Health Centre and the Cooperative Wildlife Ecology Research Network. In addition, EC's Services also funded several other chairs, grants and fellowships, for example the joint Atmospheric Environmental Service and NSERC research chairs and science subvention programs. All together, these other efforts cost the Department about $8 million in 1994/95.[16] By 1997-98, it had declined to about $5 million. When faced with the choice of preserving its own core scientific capacity or funding university research, the Department mostly chose to cut external funding. EC was not always responsible for the termination of these activities. Some joint programs with NSERC were not renewed because the granting council decided to pull its funding from them.

The budgetary cuts in the mid-1990s did not mark the end of EC's collaboration with universities. Two new networks, for example, were started in 1994: the Canadian Climate Research Network and the Atlantic Cooperative Wildlife Ecology Research Network. As usual, these arrangements tended to originate within EC's Services, often at a program or individual researcher level, and not as a Departmental initiative. In 1995, the Department decided to look at whether it could take a more strategic approach to managing these partnerships. The review was prompted by requests from both the DM for a report on EC's S&T as well as from the Minister for a one-stop window in EC for universities to access grants, contributions and contracts.[17] But science managers thought that the management of the programs and contracts ought to remain in the Services rather than be collapsed into one program.[18] The managers were most concerned about maintaining the goodwill and natural alliances that they had established with their university partners. So partnering arrangements were left where they were and EC simply established a website to better disseminate information on its funding programs.

---

[15] *Review of the Impact of the Eco-Research Program on Environmental Sciences in Canada: Lessons for the Canadian Environmental Sciences Network*, prepared by the Whetstone Group. Science Policy Branch working paper #34, May 2004

[16] This amount does not include the approximately $9 million that year for the Eco-Research Program.

[17] For further information on the report, see chapter 1. Managing partnerships more strategically was one of the challenges listed in the November 1994 *Action Plan for Managing Science and Technology at Environment Canada*.

[18] Minutes, September 1995. EC S&T Management Committee, box 2

Despite the setbacks due to budget cuts, EC still retained a strong interest in university collaborations and an ability to build on its experience when an opportunity presented itself. This happened during the election in the summer of 1997 when the Liberals made a commitment to "strengthen our environmental and health science capacity by adding $10 million annually in new funding for research on toxic substances."[19] Starting that fall, EC science policy staff together with colleagues in Health Canada, began designing what would become the Toxics Substances Research Initiative. Announced in late 1998, the Initiative ran for four years and spent about $40 million. It was focused on producing new knowledge on emerging issues on toxics that would require policy attention in the near future. What was especially innovative were its encouragement of partnerships between government and non-government researchers and its ability to provide funding to both. Most of the Initiative's projects involved both government and external scientists – usually from universities – with 60% of the total funds going to the latter.[20]

Another example (although one at a Service level) of EC's adaptive ability was the creation of the Canadian Foundation for Climate and Atmospheric Sciences in 2000. The Atmospheric Environment Service's longstanding programs of research grants to universities had been considerably undermined by budget cuts. Unable to secure new funding for these programs directly, the Service capitalized on the government's practice at that time of using year-end money to set up foundations aimed at universities.[21] Championed by the Service's ADM (Gordon McBean), the Foundation was created through a scientific association, the Canadian Meteorological and Oceanographic Society. It received $60 million in the 2000 federal budget and another $50 million in 2003.[22] The purpose of the Foundation was to build up capacity in climate research. Unlike the Toxics Substances Research Initiative, it only funded academic work. Government science was not eligible.

The Departmental S&T Management Committee also developed two policies at this time to guide partnering, which were approved by the departmental management committee. The first was prompted by a concern in the Department's legal services that employees who were adjunct professors in universities were in a conflict of interest situation.

[19] Environmental stewardship chapter in the Liberals' Red Book II, *Securing Our Future Together*

[20] *Report of the Final Evaluation of the Toxic Substances Research Initiative*, prepared by Madhu Joshi & Associates, March 2004

[21] Personal communication from Gordon McBean

[22] The Foundation's funding agreement with the federal government ended in March 2012.

The Committee developed a *Collaborative S&T Positions Policy* that recognized the value of these arrangements to the Department and encouraged its scientists and engineers to continue to seek and accept such appointments.[23] In order to protect employees from conflict of interest concerns, the policy instituted a formal approval process for appointments. The second policy was an affirmation that partnering was a key mechanism in promoting EC's vision and fulfilling its mandate. It noted the increasing importance of collaboration and its significance in enhancing the department's scientific leadership. The policy listed six principles guiding EC's S&T partnering. It was to be undertaken in the public interest; to support departmental and government-wide priorities; to enhance EC's capacity; to foster capacity building in other organizations working toward similar goals; to help build consensus among different organizations which have an impact on the environment; and, if it involved the private sector, to be carried out so as to minimize competition with that sector.[24]

Canadian Environmental Sciences Network

In late 1999, EC circulated internally a discussion paper outlining new research models to help strengthen the national system of environmental research. It suggested that the Department should organize and finance a formal system of Canadian environmental research networks.[25] The paper had been commissioned as part of the work of the Department's R&D Advisory Board looking at environmental S&T capacity. Although the Board was just completing its work in this area, the discussion paper marked the beginning of a concerted policy effort within EC to establish what would soon be called the Canadian Environmental Sciences Network.

Creating an institutional framework to bring together the many performers of environmental S&T in Canada and their major stakeholders was not a new concept in the Department. In 1989, the idea of an "Environmental Science Foundation" had been suggested in order to build partnerships among governments, universities and the private sector and to manage environmental science in Canada in an integrated

---

[23] *Collaborative S&T Positions Policy*, S&T Management Committee report #2, January 1999
[24] *Science and Technology Partnering: Principles and Practices*, S&T Management Committee report #3, February 2000
[25] *Strengthening Environmental Research in Canada: A Discussion Paper*, prepared by The Impact Group. Science Policy Branch working paper #5, December 1999

way.[26] And in 1997 and early 1998 there had been some discussions at the S&T Management and Executive committees about a national environmental sciences initiative. This would work towards ensuring that Canada would have the S&T capacity and appropriate governance structures needed to secure Canada's environmental future.[27] None of these propositions had gone very far. By 1999, however, the situation was much more supportive.

Just emerging from Program Review, EC had been turning its attention to whether it had the S&T capacity it needed (see chapter 4). At this time, the federal government was also beginning to make new investments in S&T, largely in the university sector. In the five years after 1997, the government more than doubled its expenditures for post-secondary education. Not only did the granting councils receive increases, but large amounts of year-end money were also being put into new foundations or programs such as the Canada Foundation for Innovation, Genome Canada and the Canada Research Chairs. In addition, federal S&T policy thinking was emphasizing the value of partnering. One of the principles in the 1996 Federal S&T Strategy, for example, stressed the importance of collaboration in increasing the effectiveness of government S&T.[28] As well, a report from the Council of S&T Advisors on federal S&T in late 1999 recommended that the government implement and fund new models for conducting S&T that were more horizontal, cutting across the government and the innovation system.[29] Furthermore, EC's minister was introducing at that time "a new architecture" for environmental management in Canada. One of its three pillars was active partnerships. Together, these factors provided a very favourable climate for proposals on how to mobilize and direct S&T resources for the environment.

Discussions about a Canadian environmental sciences network gained momentum in 2000. In June, a second discussion paper added to the dialogue. This one focused on university partnerships, examining opportunities for partnering that had emerged due to the new foundations and programs that the government was creating.[30] Concurrently, EC's DM (Alan Nymark) decided to appoint a Special Science Advisor, John ApSimon, to enhance relationships between the

[26] *Environmental Science Partnerships for the Future*, July 1989. EC OSA, box 1
[27] For example, *Towards a National Environmental Science Initiative: Building Environmental S&T Capacity for the New Millennium*, a presentation by Robert Slater to the CCMD Science Network, January 9, 1998. Science Policy Division papers
[28] *Science and Technology for the New Century: A Federal Strategy*, March 1996
[29] *Building Excellence in Science and Technology (BEST)*
[30] *Environment Canada University Research Partnership Expansion Strategy: A Discussion Document*, prepared by the Impact Group. Science Policy Branch working paper #12, June 2000

Department and universities.[31] The DM also decided that calling together representatives of existing environmental S&T networks would be a useful step forward in establishing a Canadian network. Such a workshop was planned in the fall of 2000 and held in Ottawa on January 26, 2001.

ApSimon chaired the workshop. In attendance were 45 individuals, representing 13 environmental networks, 4 environmental organizations, 6 universities and 4 federal departments.[32] Its purpose was to learn about the experience of these networks, to think about ways the environmental sciences in Canada could be strengthened, and to hear about the plans for a Canadian environmental sciences network and to suggest next steps. The DM gave a presentation outlining the importance of the environmental sciences, EC's priorities, and what it would take to create a Canadian network. Participants were generally interested in the idea, but wanted further details on it before buying in.[33] They also found the meeting to be a very useful opportunity to meet, as they did not normally interact with one another.

EC's approach to a Canadian Environmental Sciences Network was to establish it as an independent entity with its own secretariat and board of directors. That would require funding. The Department requested $12 million through a memorandum to cabinet in the spring in order to put an organization in place that would further scope out and champion the Network. While waiting for a decision, EC set up a website for the Network, continued to develop communications material and consulted within the federal government and with universities. The Department's vision was that the Network would be the hub supporting and strengthening existing environmental S&T networks. It would provide a national focal point for the environmental sciences in Canada. The Network's goals were to better connect those who create, apply and fund environmental knowledge; to help environmental scientists to collaborate across the boundaries of discipline and geography; and, to champion the role of environmental knowledge in Canada. Aside from fostering linkages between scientific networks and between providers and users of knowledge, EC saw the Network developing a national agenda and a rationalized investment strategy for the environmental sciences.

---

[31] ApSimon had just retired from Carleton University as its vice-president for research and external relations. He started working at EC in September on a part-time basis and left in March 2004.

[32] EC was involved in each of the 13 networks.

[33] A full report on the workshop can be found in *Canadian Environmental Sciences Network (CESN): Discussion Paper.* Science Policy Branch working paper #22, March 2001

Funding for the Canadian Environmental Sciences Network did not come through. By the fall of 2001, an alternative plan for moving forward had been developed. Based on input from the S&T Advisory Board and considerable discussion in the Department, the plan focused on working towards a national research agenda for the environmental sciences by working with the granting councils and foundations, and on nurturing regional networks within EC. ApSimon took the lead in working with the councils and foundations. Over the next two years, several initiatives were undertaken. There was work with NSERC to assess the environmental science capacity in universities. A joint workshop was organized with the CFI to explore how it might do more to support environmental research. Meetings or workshops were also held with the CIHR and Genome Canada. And a series of workshops were held to provide input into future SSHRC initiatives in the environment. Working with the granting councils, EC also produced a survey of environmental science networks in Canada, finding 257 of them – 70% of which were research networks.[34] While useful activities in getting these bodies to focus on the environment, they do not appear to have led to any new programs. Indeed, it is probably fair to suggest that their interest was largely in the possibility of obtaining increased budgets. More generally, it was clear to those involved that developing a national environmental R&D agenda was an enormous task. Broad consultations would be required, especially given the very disparate nature of the environmental sciences community.[35]

EC's science management committees, with ApSimon's participation, worked on the second thrust of the alternative plan, which was to look at establishing environmental sciences networks in each of the Department's five regions. The only successful effort was in the Atlantic region. There, EC staff built on the Atlantic Cooperative Wildlife Ecology Research Network to create the Atlantic Environmental Sciences Network. Having consulted with Deans of Arts and Science in Atlantic Canada universities in the fall of 2001, EC organized a two-day workshop in May 2001 to launch the Network. It involved universities, provincial and federal departments, industry, NGOs and first nations. Like the proposed Canadian Environmental Sciences Network, it was set up as a network of cooperative groups organized around six themes. Among its

[34] *A Survey of Environmental Research Networks and Partnerships*, prepared by Nicole Bégin-Heick & Associates. Science Policy Branch working paper #35, March 2003

[35] *National Environmental R&D Agenda Setting: A Commentary on Issues, Options and Constraints*, by Bruce Doern & Michael Rosenblatt. Science Policy Branch working paper #14, March 2001

major purposes were the promotion of research to address environmental issues in Atlantic Canada and to facilitate linkages among its members.

Although the effort to establish a Canadian Environmental Sciences Network could claim a number of successes in 2002 and 2003, the initiative was losing the interest of senior management. ApSimon's contract was not renewed beyond March 2004. And fewer policy resources were devoted to the issue. Yet, the Network concept did see a revival in the summer of 2004 when the Department began work on its new Competitiveness and Environmental Sustainability Framework. Championed by the DM (Samy Watson), the Framework consisted of five national pillars, one of which was S&T. The ideas that had been developed under the banner of the Network were adapted for the S&T pillar. They included a national environmental S&T agenda, coordinating council and funding mechanism. For about a year there was considerable policy work on the Framework and its pillars, but it did not get the buy-in from other parties that had been hoped for and began to fade away as a Departmental priority. By the spring of 2006, the DM directed that work under the S&T Pillar should focus on departmental S&T management and the integration of federal S&T rather than on national S&T issues.

R&D Facilities at Universities

One type of collaborative arrangement with universities that drew considerable policy attention over many years was locating R&D facilities on university campuses. EC had gained a fair amount of experience with this. The National Hydrology Research Centre had been moved from Ottawa to the University of Saskatchewan in 1986. The Canadian Centre for Climate Modelling and Analysis was relocated from Downsview to the University of Victoria in 1993. This was done to establish closer links between EC's climate modellers and the ocean modellers at the University and at the Department of Fisheries and Oceans' Institute of Ocean Sciences also near Victoria.[36] And the National Wildlife Research Centre was moved from Gatineau to Carleton University in 2002. That move was prompted by the need to construct a new building to replace an aged facility plagued with health and safety concerns. But the ADM of the Environmental Conservation Service (Karen Brown) used this need to push for the location of the building on a university campus in order

---

[36] *Smart Partners: Innovations in Environment Canada – University Research Relationships*. Science Policy Branch working paper #33, 2004

to develop synergies in wildlife research.[37] After an exploration of possibilities, Carleton University was chosen.

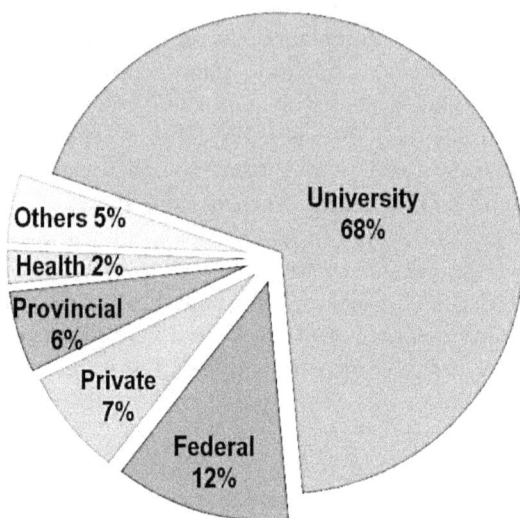

Figure 8.1 Distribution of EC bilateral scientific collaborations by sector (2003–2007)[38]

EC was not alone among federal departments and agencies in having a history of co-locating R&D facilities. It was surprising then that the government's March 2007 budget included a commitment to transfer some laboratories.

> The Government will launch an independent expert panel that will consider options for transferring federal laboratories to universities or the private sector. The panel will report to the President of the Treasury Board in the fall of 2007 on the type of non-regulatory science that should be transferred, which partners should be involved and an appropriate governance framework. The panel will also be asked to identify up to five laboratories that could be early candidates for transfer.

---

[37] Ibid.

[38] As measured by authorship of peer-reviewed published articles. *A Bibliometric Analysis of R&D at Environment Canada.* PowerPoint presentation, Science-Metrix, 2009

It was disconcerting because it seemed out of place for many science policy analysts, appearing as a throw back to the motives behind the former contracting-out policy. Now that university research was well funded, thinking in policy for science had shifted to finding ways to link the capacity in the various sectors of the Canadian science and innovation system. Just a few years earlier, for instance, the Council of Science and Technology Advisors had recommended that the government "embrace a vision of a linked S&T system, championing S&T collaboration as a core way of doing business."[39] The Council stressed that each of the three major S&T sectors – university, government and industry – had to be strong players in their own right and be prepared to contribute substantively to a linked system.

The independent panel was set up in August and submitted its report in early 2008. Significantly, the panel's report chose to reframe transfer, shifting from divestment to inter-sectoral partnerships.[40] It also developed a framework and nine criteria for assessing proposals for enhanced collaboration. In keeping with its mandate, the panel did recommend five federal laboratories as early candidates for "new management arrangements." One of these was EC's Wastewater Technology Centre, which had been the subject of an unsuccessful government-owned, contractor-operated experiment between 1991 and 1999. But, perhaps acknowledging that policy had moved beyond simply outsourcing, the government responded to the report by selecting only two of the labs. NRCan's Geoscience Laboratories and AAFC's Cereal Research Centre were requested to develop business cases for further study.[41] Treasury Board withdrew from any facilitating or coordinating role, leaving it up to departments to deal with their S&T facilities as they saw fit.

---

[39] *Linkages in the National Knowledge System (LINKS): Fostering a Linked Federal S&T Enterprise*, 2005

[40] *Inter-Sectoral Partnerships for Non-Regulatory Federal Laboratories*, report to the President of the Treasury Board of Canada. The report was released to the public in June 2008.

[41] Both labs are still parts of their departments.

9

Contributing to Industrial Innovation

EC was created due to concerns about the quality of the environment; its core mission was not focused on supporting economic development. However, the two areas are closely intertwined. So, despite its primary focus, the Department has been involved with industrial issues throughout its history, particularly but not exclusively through its regulatory role.

The concentration of federal S&T policy on stimulating industrial innovation created a dilemma for EC. The government's various attempts at strengthening the competitiveness of Canadian firms threatened to displace the goals of EC's S&T. The Department struggled to remind the government about the other roles of federal S&T and to preserve its environmental focus. At the same time, the government's efforts also provided opportunities for EC to advance that part of its mission related to protection of the environment. Federal funding for developing and promoting new environmental technologies could help industry comply with regulations developed by the Department for improving the quality of the environment.

EC's policy for science was engaged with industrial innovation from the start. Reflecting the evolution of federal policy, EC's moved from assisting industrial R&D, to transferring research and technology, to fostering environmental technology and environmental industry. While never a large part of its activities, EC's policy work related to industry was not insignificant.

Despite its activity, EC did not wholeheartedly embrace involvement in industrial innovation. For example, interest in supporting environmental industry varied greatly over the years, from a high level of activity in the late 1980s to rejection of that role twenty years later. That change was likely due to the reduction of funding available to the Department following Program Review, and the consequent focusing on what were generally agreed to be EC's core activities. Since several other federal departments – ones with more direct economic mandates than EC's – were actively pursuing a role in environmental technologies, it was easy to cede the field to them. In addition, a more adamant emphasis on the regulatory role of the Department in the last decade made a defence of a technology promotion role difficult.

Given the Department's ambivalence regarding its role in industrial innovation, it has had a difficult time in defining its policy for technology and greater difficulties in implementing that policy. However, driven by the notion of sustainable development, environmental technology has persisted as an important consideration in EC's policy efforts. The 2009 *Technology Role* captures the Department's current view of its role with respect to environmental technology. But the past experience of EC's engagement in industrial innovation suggests that the specifics of that agreement are unlikely to endure.

Research and Technology Transfer

The decade prior to the establishment of EC saw the creation of a number of major federal programs aimed at stimulating R&D in Canadian industry.[1] When the Department opened its doors in 1971, these programs were spending about $100 million a year. While much of government policy was directed to strengthening Canadian industry, this was not EC's priority. Although the Department was one of the federal government's largest spenders on R&D, a mere 2% of that went to industry.[2] Its original six goals made no mention of industry. However, the first goal was to carry on the programs of the units that had been brought together to form EC. Fisheries and Forestry did provide support and services to their natural resources sectors and, to a limited extent, to some manufacturers as well. For example, Forestry funded a research program on pollution abatement in the pulp and paper industry, which was administered by a joint government-industry committee.[3]

In addition, another of EC's goals, to clean up and control pollution, created a niche for the promotion of environmental technologies. In 1973, for instance, cabinet approved the Department's cost shared program, Development and Demonstration of Pollution Abatement Technology (DPAT), aimed at controlling air and water pollution.[4] It also created the Development and Demonstration of Resource and Energy Conservation Technology program in 1978. And the Office of the Science Advisor's Advanced Concepts Centre wrote a report on

---

[1] Andrew H. Wilson, *Governments and Innovation*, Science Council of Canada Background Study no. 26, April 1973

[2] Memo on R&D expenditures 1971-72. LAC Acc. 1993-94/003, box 9

[3] Douglas Jones, *A Review and Index of the Cooperative Pollution Abatement Research (CPAR) Program of the Federal Government and the Canadian Pulp and Paper Industry, 1970-1979* (Fisheries & Environment Canada, 1980)

[4] Both DPAT and CPAR were cancelled at the end of the 1970s due to budget cutbacks.

environmentally appropriate technology that became a best seller in 1975 for the Department of Supply and Services.[5]

Soon after its creation, EC was swept up in a new federal approach to promoting industrial innovation. The government introduced the Make-or-Buy Policy in 1972.[6] It required departments to contract out more of their R&D. It was designed to use the government's purchasing power to induce more R&D capacity in industry. EC devoted considerable effort to implementing the Policy. It led to a quick increase in EC's contracting out to industry; by 1975-76 those contracts had quadrupled in value. But their total amount remained a small part of EC's R&D expenditures. Moreover, EC's contracts had little impact on strengthening R&D in manufacturing, the Policy's target. An analysis done for the Economic Council of Canada in 1982 concluded that no significant benefits from the Policy should have been expected of EC.[7]

The ongoing preoccupation with the country's weakness in industrial innovation gave rise to yet another federal initiative in 1978. Based on several studies by the Science Council of Canada, the government decided that all federal labs were to ensure that the technology they developed would be transferred to industry.[8] In addition, the concept behind NRC's Pilot Industry/Laboratory Program, which had been set up in 1975 to transfer research results from the NRC to firms, was extended to some other departments, including EC. This became known as the Cooperative Projects with Industry program. Under it, EC received $400,000 to promote the commercial development of instruments and machines designed in its labs.[9] The program led to the development of an interdepartmental community of expertise in technology transfer, which became the Interdepartmental Group on Intellectual Property Management and later Federal Partners in Technology Transfer.

---

[5] Supply and Services sold all of their 1500 copies of Bruce McCallum, *Environmentally Appropriate Technology: Developing Technologies for a Conserver Society in Canada*, March 1975. LAC Acc. 1993-94/003, box 4

[6] The Policy is discussed in greater detail in chapter 4.

[7] A. B. Supapol & D. G. McFetridge, *An Analysis of the Federal Make-or-Buy Policy*, discussion paper no. 217 (Ottawa: Economic Council of Canada, June 1982)

[8] In particular, the Council's report no. 24, *Technology Transfer: Government Laboratories to Manufacturing Industry*, December 1975, and its background study no. 35, *The Role and Function of Government Laboratories and the Transfer of Technology to the Manufacturing Sector*, April 1976, by Arthur J. Cordell and James Gilmour

[9] The Cooperative Projects with Industry program was merged into PILP in 1981, which in turn became an element of the NRC's IRAP in 1986. That part of IRAP was closed in 1990 as part of NRC budget cuts. William M. Coderre, *Inside IRAP 1962-1992* (Blurb. com, 2011)

In adherence to the government's decision, EC prepared a policy directive on technology transfer in December 1978.[10]   Lab heads were responsible for making their technology available to industry, for deciding on how best to do this, and for keeping records of the transfers. ADMs were to ensure compliance with the directive. However, the technology transfer policy, like Make-or-Buy, had a modest impact on EC. Most of its research centres were, as Roots noted, "not in general concerned with the kind of technology that has a product manufacturing or commercial service interest."[11]   Their main function was to serve departmental needs.

The low priority EC normally gave to industrially relevant work was very evident during times of budgetary restraint. The Forest Products Labs were among the few in the Department that had transferring technology as part of their objectives. The 1978 budget cuts to EC resulted in a decision to privatize the Labs, justified in part as a "withdrawal from matters that can be done well by industry."[12]   This was not the first time that research capacity in Forestry had been transferred. In 1975, its logging research had been combined with that of the Pulp and Paper Research Institute of Canada to form the non-profit Forest Engineering Research Institute of Canada.

The separation of Fisheries from EC in 1979 and of Forestry in 1984 left very little industrially oriented R&D in the Department. The most prominent of what remained was the Wastewater Technology Centre. During the 1980s, several studies had recommended that the government make use of the GoCo (government-owned, contractor-operated) model, which had been in operation for many years in the United States. Faced with tight budgets and a desire to improve the commercialization of its technologies, the Environmental Protection Service decided to transfer the management of the Wastewater Technology Centre to a private company, RockCliffe Research & Technology. After extensive negotiations in 1990, a three-year agreement was reached in 1991. RockCliffe would manage the Centre, which would continue to be owned by the government. Its purpose was to increase the Centre's contacts with the private sector and to more quickly and effectively commercialize its research. This was a novel experiment in Canada at the time; but one that was not replicated elsewhere. The arrangement proved only moderately successful and was extended, under several contractors, until 1999 when

---

10 "DOE's Policy Directive on Technology Transfer", in *Technology Transfer from Federal Laboratories to Industry*, MOSST, May 1980. EC Roots, box 3
11 Memo, April 1978. LAC Acc. 1993-94/003, box 29
12 Memo, August-September 1978. LAC Acc. 1985-86/342, box 5. FORINTEK was created in 1979 to take over the labs, and was funded in part by the federal government.

EC once again took over responsibility for the management of the Centre.[13]

Environmental Industry and Environmental Technology

As the Department concentrated on its environmental mission after the departure of Fisheries in 1979, a new dimension in its relationship with industrial innovation opened up, one that focused on environmental industries and environmental technologies. EC pursued various opportunities throughout the 1980s and early 1990s to increase support for these areas and to promote their needs in discussions about federal S&T policy.

Seeking to better leverage the government's R&D spending on industry, the Ministry of State for Economic Development asked departments in 1980 to prepare reports on 13 industry sectors. One of these was the environment sector. EC's report, *A Plan for Federal Encouragement of Industrial and Industry-Oriented R&D in the Environment Sector*, was prepared during the summer of 1980. It noted that approximately 77% of the S&T contracts awarded by the Department of Supply and Services for EC, worth more than $17 million, went to performers in the environment sector.[14] The report identified 11 opportunities for growth in the level of R&D conducted by or for industry: environmental monitoring, space applications, environmental planning, energy conservation, hydraulic power generation, solar and wind energy, waste management, management of in-situ water quality, weather modification, communications and information systems, and transportation. To realize the potential for growth in these areas, EC asked for $75 million and 100 person years.

Budget cuts in 1978 had resulted in the closure of EC's research and technology programs for pollution abatement. An attempt to fund these areas through the Enterprise Development Program of Industry, Trade and Commerce had not proven satisfactory. This prompted John Roberts, then Minister of both MOSST and EC, to establish a task force on environmental protection technologies in August 1981. It was asked to consider the economic benefits to the country, both for the optimal

---

[13] For further information on the RockCliffe form of GoCo see *Models and Policies for the Commercialization of Government Science and Technology*, prepared for ISTC, August 1989. For an overview of the results of the WTC experiment see *Future Options for Environment Canada With Respect to the Operation of the Wastewater Technology Center*, prepared by Apogee / Hagler Bailly for EC, September 1998.

[14] *A Plan for Federal Encouragement of Industrial and Industry-Oriented R&D in the Environment Sector*. EC Roots, box 3

economic exploitation of its natural resources and for growing a Canadian industry to take advantage of a worldwide market for these emerging technologies.[15] The task force's report, in February 1983, called for the federal government to play a major role in the development of these technologies and made several recommendations to that effect.

Although EC was actively seeking support for the environmental industry sector, the Department usually did not consider it to be a key priority. When the Task Force on Federal Policies and Programs for Technology Development reported in July 1984, it recommended making that function "an explicit part of all appropriate departments' mandates." EC was quick to remind MOSST of the other significant roles of federal S&T.

> There is another matter I want to bring to your attention. As I noted in my recent letter to you, we are concerned that the Wright Task Force acknowledged and then lost the distinction between federal laboratories in support of industry and those which serve other important government goals. It appears that this discussion document also has failed to grasp the significance of science activities in this department and others which are devoted to the preservation and enhancement of the human health-environment-economy complex. Our work is essential if we are to preserve for Canadians the quality of life in which they can enjoy the benefits of their economic activities. ... It therefore strikes us as very short-sighted to propose that the roles of government laboratories be redefined so that they are more consistent with the objectives of economic growth, narrowly defined in terms of the manufacturing sector. We have other important work to do.[16]

The following year, MOSST launched a review of technology centres as a follow up to the Nielsen Task Force. Two of EC's research centres, the Wastewater Technology Centre and the River Road Environmental Technology Centre, were included on a list of industry-oriented labs that would be required to recover more of the costs of their activities from industry with the goal of making them more client oriented. EC did not believe that the two centres should have been included, arguing that their

---

[15] *Report of the Task Force on Environmental Protection Technologies to the Minister of State for Science and Technology*, February 1983
[16] Letter from the ADM of the Corporate Planning Group, December 1984. EC Roots, box 8

primary role was in support of public policy. It worked hard over the next year to remove them from the list.[17]

Despite EC's reservations, it was impossible for the Department at that time to ignore engagement in industrial innovation. The issue of integrating economic and environmental decision-making had a very high profile. The United Nations Brundtland Commission visited Canada in May 1986, popularizing the notion of sustainable development.[18] In its wake, the Canadian Council of Resource and Environment Ministers established a national task force on the environment and the economy. In addition, Canadian S&T policy continued to be strongly preoccupied with the challenge of increasing industrial innovation in the country.

EC worked to more systematically define the environmental protection industry.[19] It also attempted to build on the 1987 federal S&T strategy, *InnovAction*, to expand to other universities the industrial research chairs and technology centres in environmental systems engineering it had set up at McMaster University and those in pulp and paper wastewater treatment at the University of British Columbia. Although the effort was unsuccessful, the Department did not give up. At the same time it also sought $50 million for a five-year fund for an environmental technology innovation program. These funds would be used to promote upgrading of existing technologies and the enhancement of private sector skills, technology transfer and demonstration of new technologies in areas of interest to the Department. This proposal was also not funded. However, its ideas would soon find more fertile ground in the context of the Green Plan.

Technology for Solutions was the main initiative under the Green Plan aimed at accelerating the development, transfer, commercialization and use of advanced environmental technologies. EC's Minister and the Minister of Science jointly announced the $100 million program in October 1991. It had three components. The largest was an $80 million environmental technology commercialization program. Its focus was on demonstration projects for new environmental technologies in EC's areas of interest: clean process technologies, waste reduction and recycling, air and water pollution control and water conservation. The program provided repayable loans of up to 50% of eligible costs. Industry, Science and Technology Canada administered it, with EC participating on the

---

[17] EC Roots, box 4
[18] The report of the UN World Commission on Environment and Development was published in 1987.
[19] *Inventory of Canadian Environmental Industries*, February 22 1988

steering committee.[20] It is difficult to determine how much of the program's funding survived the cuts made during the life of the Green Plan or the Program Review period.

The two other components, a technology transfer program and an environmental technology network, were administered by EC's Technology Development Branch. The former aimed at providing services to firms to help them locate, assess, transfer and promote environmental technologies. Following a recommendation from a study by Denzil Doyle on an appropriate infrastructure for delivering these services, EC funded the establishment in 1994 of three environmental technology advancement centres. They operated at arm's length from government, and were located in Alberta, Ontario and Quebec.[21] The purpose of the network program was to link existing government and university centres of environmental technology in order to promote increased awareness of the latest R&D in the field and to increase communication and cooperation among researchers. Apparently because of cuts to the Green Plan this component never developed into a separate activity, becoming a function of the three centres.

Despite these various activities, work on environmental technologies remained a minor part of the Department's total effort. A 1992 overview estimated that about 20% of EC's R&D effort was devoted to the development of various environmental technologies, with the major focus on waste water treatment, waste disposal, site remediation, oil spill detection and clean-up, and air pollution measurement.[22] This work and its funding were concentrated in the environmental protection components of the Department. The result was that policy for environmental technology was mostly done there. Policy for science and policy for technology in EC were generally conducted in isolation of one another.

---

[20] EC had originally planned to be responsible for the entire Technology for Solutions program. But Industry, Science and Technology Canada suggested that it be the lead. DM discussions resulted in that Department being the lead on the commercialization fund with project approval being made by both departments. Memo to the DM, June 11 1991. EC OSA, box 2

[21] D. J. Doyle *Building a Stronger Environmental Technology Exploitation Capability in Canada*, prepared for EC and ISTC, July 1992. The three centres – the Ontario Centre for Environmental Technology Advancement, the Canadian Environmental Technology Advancement Corporation - West, and Enviro-access – are still operating. For further information see Bert Backman-Beharry and Robert Slater, "Commercializing Technologies Through Collaborative Networks: the Environmental Industry and the Role of CETACS" in *Innovation, Science, Environment: Canadian Policies and Performance, 2006-2007*, ed. G. Bruce Doern (Montreal: McGill-Queen's University Press, 2006), 169-193

[22] *A Compendium of R&D in Environment Canada*, prepared by the Office of the Science Advisor, May 1992

EC was not the only science-based department involved with environmental technologies. Energy, Mines and Resources and the National Research Council, for example, were significantly involved in areas related to their mandates. And other federal departments who were not performers of S&T, such as CIDA and External Affairs and International Trade, were also influenced by the same forces affecting EC and played significant roles in promoting environmental technologies. Industry, Science and Technology Canada identified a role for itself in developing and applying innovative solutions to environmental problems. Through its environmental industry sector campaign, which started up in 1989, that Department sought to foster positive investment climates, the development of new and more effective technologies, and the growth of the Canadian environmental industry.[23] That work brought it into close contact with EC. For example, that Department administered an $18 million environmental technology development program on water pollution abatement as part of the St. Lawrence River Action Plan, which was led by EC. The overlap of interests amongst many departments and the consequent need to negotiate respective roles would be a constant feature in EC's policy work on environmental technology.

Promoting Technology versus Regulation

During the federal election campaign of 1993 the Liberals announced an environmental industry initiative, promising that 25% of all new R&D funding would be devoted to environmental technologies. Following the Liberal victory and after several months of consultations, the Canadian Environmental Industries Strategy was announced in September 1994. Responding to criticisms that existing federal programs were fragmented and uncoordinated, the Strategy established a steering committee whose members were drawn from environmental companies, relevant associations, the provinces and the federal government.[24] The committee was co-chaired by EC and Industry Canada.

The Strategy built on existing or already planned departmental activities, and was boosted by another $57.5 million in new funding. The centrepiece of the additional funding was the $33.5 million environmental

---

[23] Environmental Industries Directorate, Environmental Affairs Branch, Industry, Science and Technology Canada, *Environmental Industry Sector Initiative. Presentation to Departmental Management Committee*, March 18 1992

[24] The 1993 *Inventory of federal environmental industry programs and activities*, compiled by EC, listed 82 programs. See also Anne Fouillard, *Emerging Trends and Issues in Canada's Environmental Industry*, working paper no. 8, National Roundtable on the Environment and the Economy, and the Institute for Research on Public Policy, March 1993

technology development and demonstration initiative. Like the Green Plan's commercialization program, the initiative was aimed at assisting firms in demonstrating new environmental technologies. The initiative also felt the impact of Program Review. It was frozen in the February 1995 budget. However, it was resurrected the following year as the environmental technology component of the new Technology Partnerships Canada.[25]

The Strategy marked a watershed in EC's policy work on environmental technology. For over five years the environmental protection side of the Department had been positioning itself to support the growing environmental industry sector.

> In developing our new model for R&D Strategy our thinking went far beyond the traditional role of government to regulate industry in order to reduce the stressors on the environment. Our fundamental goal is to see achieved in Canada a state of sustained utilization and sustained benefits from natural resources. In support of Canadian industrial activity, the agriculture, fisheries and forestry sectors, tourism and trade, we wish to harness our R&D capacity to optimize environmental and economic benefits. We believe we can contribute to enhancement of resource management for sustained use and benefits for Canadians.
>
> As well, C&P will place increasing emphasis on promoting the development, transfer, and commercialization of improved technologies for conserving and protecting the environment, in order to provide environmental and economic benefits.[26]

Following the Strategy, the focus on environmental industries began to wane, replaced by an emphasis on technology for environmental priorities.

The reasons for this shift appear to be mostly due to Program Review and to a more cautious approach to the regulatory function of the Department. Program Review made deep cuts in the funding for all EC programs, including those for assisting firms. When new funding for technology programs did again become available, it was often placed under the direction of non-governmental foundations. A good example of this is the Canada Foundation for Sustainable Development Technology, which was announced in the federal budget of 2000. The Foundation first received $100 million to fund the development and

---

[25] This program ran until the end of 2006.
[26] Conservation and Protection Service, *R&D Prospectus 1993-1994*. EC OSA, box 3

demonstration of sustainable development technologies related to climate change and clean air. It subsequently was given further funding and an expanded mandate including clean water and soil and next generation biofuels.[27]

In addition to the lack of money for environmental technologies, which the Department could apply to its own agenda, Program Review also forced the Department to make difficult choices about its core business. Even within environmental protection in EC, doubts began to be expressed about the appropriateness of the Department being both a regulator and a promoter of environmental industry. The lack of a champion for environmental technology within the Department also led to the movement of key personnel to other areas and other departments.

The locus for Departmental activity in environmental technology was in the Environmental Protection Service, which had been re-created in the 1993 reorganization of EC. The Service's Environmental Technology Advancement Directorate was the home to groups working on clean technologies, environmental industries and technology transfer.[28] That work slowly evolved in the second half of the 1990s away from environmental industry and the promotion of environmental technologies. Instead, the groups devoted their energies to liaison and administration functions. For example, they provided oversight for a number of external groups involved in environmental technology such as Sustainable Development Technology Canada, the Green Municipal Fund, the three environmental technology advancement centres, and the environmental technology verification program.[29] They were also engaged in related federal programs, for instance the Climate Change Action Plan and the Technology Early Action Measures program. Finally, they represented the Department's environmental interests in federal technology investment programs or initiatives like the Canadian Biotechnology Strategy, and sometimes administered special funding for research on the environmental dimensions of emerging technologies such as biotechnology and biofuels.

The Directorate also housed the Department's intellectual property office. Following the closure of Canadian Patents and Development Limited in 1991, intellectual property assets were transferred back to the departments that had produced them. EC was never a major federal

---

[27] Federal funds received by the Foundation now total over $1 billion.

[28] The Directorate also included the Environmental Technology Centre and was responsible for the Wastewater Technology Centre. In addition, later in the 90s, it became responsible for programs and policy development on municipal wastewater and contaminated sites.

[29] Work on environmental technology verification, which continues to this day, had its origins as one of the initiatives of the 1994 Canadian Environmental Industry Strategy.

player in patents, but did occasionally develop some valuable intellectual property – for example, the Brewer Spectrophotometer, the Microwave-Assisted Process, and the Forecast Production Assistant.[30] EC did not have the resources to actively identify which of its own environmental technologies had commercial potential or to fully engage in transfer efforts. As was the case with most other public bodies, the revenue from EC's intellectual property was insufficient to cover its management costs.[31]

The Technology Role

The broad range of environmental technology initiatives with which EC was engaged and the persistent questioning about the Department's role in this area led to calls in 2003 for a more strategic approach. For example, a presentation on making the case for EC's S&T, given at the Departmental management committee retreat in March, suggested that an environmental technology strategy be developed.[32] This internal movement was reinforced by political developments. Paul Martin, campaigning to be the next leader of the Liberal Party and hence succeeding Jean Chrétien as Prime Minister, had been speaking about new approaches to technology commercialization and the need to increase investment in environmental technologies.[33] With Martin in power, these ideas were reiterated in the February 2004 speech from the throne. Within EC, the Environmental Technology Advancement Directorate worked with the corporate policy group to produce a draft strategy, "Towards a New Federal Strategy for Canadian Environmental Technologies."[34]

The federal budget in 2004 did commit money for environmental technology, but none of it for EC's agenda. The policy work, however, continued under a new initiative, the Competitiveness and Environmental Sustainability Framework. This was an effort by EC's new DM (Samy Watson) to produce a comprehensive, national framework to guide work for sustainable development. It would include shared national environmental objectives, and be developed in collaboration among

---

[30] EC's total revenue from licenses for the period from 1995 to 2008 was about $8 million. Communication from Mike Landreville, EC Intellectual Property Office

[31] For further information on the challenges of managing EC's intellectual property, see *Evaluation of Intellectual Property Management. Final Report*, EC Audit & Evaluation, July 2006.

[32] Meeting, May 2 2003. EC S&T Management Committee, box 5

[33] See for example, his speech, "Building the 21st Century Economy", given to the Montreal Board of Trade, September 18 2003.

[34] Meeting, March 24 2004. EC S&T Management Committee, box 5

federal and provincial/territorial governments, industry, and other stakeholders. The Framework's purpose was "to attain the highest level of environmental quality as a means to enhance the well-being of Canadians, preserve our natural environment, and advance our long-term competitiveness."

The Framework included five supporting "pillars," one of which was S&T. It was here that EC's work on an environmental technology strategy continued, along with that on a Canadian environmental sciences network.[35] During a presentation on the S&T pillar in July 2005 to the Department's senior managers, the DM requested a rationalization of the role of EC in environmental technologies as well as an inventory of the Department's work in that area. A working group was established to carry this out with the additional intention of developing a strategic plan. It would be several years until the work was brought to a conclusion.

A major reorganization of the Department took place in the fall of 2005. The Environmental Technology Advancement Directorate was dismantled. Its S&T components were moved to the new S&T Branch.[36] The units that had been primarily engaged in liaison functions with respect to environmental technologies became the Technology Strategies Division within the S&T Strategies Directorate. After many years apart, policy for technology and policy for science were again under the same DG. The work to explain EC's role in environmental technology continued to be led by the Division. However, it was somewhat in the shadows as the Branch focused on another request by the DM, a science strategy. Although there was some consideration given to preparing an S&T Strategy, the Branch decided to move forward without the technology dimension. The rationale was that a technology strategy would be prepared when there was consensus on EC's role in that area.

The work on EC's role in technology progressed in fits and starts over the next three years. In October 2006, a major workshop was held in Montreal to develop strategic directions for the Department's activities related to technology.[37] The following summer a draft strategy was prepared which was closely modeled on the organization of the *Science Plan*. It took another year and a half until EC's technology role was finally accepted by the departmental management committee, in February 2009.[38] The main stumbling block had been the role of the Department

[35] For details on the network see chapter 8.

[36] The Intellectual Property Office was moved to the Finance and Corporate Branch.

[37] *Developing a Technology Plan for Environment Canada: Workshop Report*, prepared by Stratos, October 20 2006. Science Policy Division files

[38] *Environment Canada's Technology Role: Towards Finalizing EC's S&T Plan*, presentation to Environment Management Committee, February 25 2009. Science Policy Division files

in providing direct assistance to industry to support technology deployment or commercialization. In addition, there were concerns about the development of technologies that were not in support of EC's priorities. The Department decided that it would not engage in such activities. They were better addressed, it believed, by other federal departments.

The departmental management committee saw EC's role in environmental technology as one of generating knowledge on the environmental impacts of technologies in order to support policy and regulatory development; providing oversight of third-party, federally-funded technology initiatives; and, developing performance criteria and measures to help define the environmental dimension of federal technology investment programs. Environmental technology became a visible part of the reporting structure for EC's programs and activities for the year 2010-2011. Reflecting its recent history in the Department, environmental technology was listed among a set of sub-activities under climate change and clean air, contributing to the strategic outcome of a clean environment.[39]

---

[39] *EC Program Activity Architecture 2010-2011*

PART FIVE

COMMUNICATIONS

10

Communicating S&T

Communicating S&T involves many objectives. Sharing advances in knowledge with other scientists is one. Publication of research results has always been a major activity in EC, but a form of communications that was largely taken for granted and hence not the subject of much policy attention.[1] Another objective is communicating scientific information to the general public. The Department has long been engaged in informing the populace about environmental issues such as acid rain, ozone depletion and climate change. These activities often included scientific and technical information. Some of them – Hinterland Who's Who, weather bulletins and David Phillips's weather trivia – have become well-known elements of Canadian popular culture. This kind of communications too was rarely the subject of departmental policy.

EC policy work on communicating S&T tended to focus on another objective, one aimed at explaining the role and value of EC's S&T to opinion leaders, central agencies, the media and the general public. This work was usually stimulated by external pressures such as criticism of the management of science, requests for reporting, budgetary cuts and, sometimes, negative publicity. It was pursued in order to correct what was perceived, by many in the Department, to be a widespread lack of understanding of the functions of EC's S&T and of awareness of its contributions.

EC's senior management became especially concerned with telling the Department's S&T story after the experience with Program Review in the 1990s. The Department, on its own and through interdepartmental collaboration, conducted S&T communications activities, many of them very innovative. Successful implementation usually depended on involving S&T employees and on working in tandem with the Department's communications staff. However, tension also arose from the differing operating cultures of the two groups. Friction between the values of open dialogue and of controlled messaging sometimes produced their own S&T communications issues.

Developing action plans, frameworks and strategies for communicating EC's S&T turned out to be relatively easy to do. The

---

[1] See Appendix 3 for EC's output of scientific articles in peer-reviewed journals.

challenge lay in funding and in providing ongoing commitment to carry out these policies.

Public Awareness

EC's first decade saw very little departmental policy work on communicating its S&T. The issue rarely came up at the departmental management committee. In 1974 the committee approved participation in the NRC's scientific and technical information system and referral service, and asked for a review of the issues involved in publication of S&T material in both official languages.[2] And, in 1975-76, it looked at the relationship between the Department's information services and S&T.[3] Disseminating S&T was left up to the individual Services.

The situation changed in 1979 when the management of EC's science came under widespread criticism. The ensuing Science Review resulted in 26 areas of concern being acted upon by the departmental management committee. Among them were several dealing with communications. Services were asked to increase public awareness of their scientific activities, to ensure that copies of scientific reports were archived in the departmental library, to give more publicity to scientific publications, and to provide public information summaries of them. The Information Directorate was also drawn in. It was to consult with the Services about providing assistance in communicating S&T information.[4]

During the next couple of years both the Information Directorate and the Corporate Planning Group attempted to draft policies regarding S&T publishing, but neither effort was successful. In 1983, Roots suggested to the departmental management committee a "modest but deliberate program" to enhance the public role of EC's science.[5] His ideas included publicizing success stories, preparing educational materials incorporating recent EC science, and encouraging scientists to participate in information activities. The committee asked for the development of a departmental policy on S&T publications.

At about the same time, Roots had been tasked with preparing a departmental policy on science.[6] The *Policy Respecting Science and Technology* was ready by July 1984 and approved by the departmental management

---

[2] Meetings, October 24, December 12 1974. LAC Acc. 1991-92/017, box 11

[3] Meetings, July 17, October 16 1975, January 8 1976. LAC Acc. 1991-92/017, boxes 14 & 15

[4] Special meeting, September 16 1980. LAC Acc. 1991-92/017, box 35

[5] Minutes of the meeting of the Senior Management Committee, November 16 1983. EC Roots, box 7

[6] See chapter 2.

committee that October. It listed communications – EC will "communicate forcefully, in a form that can be understood by and is available to government policy-makers, industry and the general public" – as one of the main functions of the Department's S&T. Roots also reminded the committee of its earlier request for a policy on S&T publications. However, that policy does not appear to have been developed. Like the *Policy Respecting Science and Technology*, it was soon forgotten as the Department struggled with pressures on its S&T capacity and other priorities.

Positioning S&T after Program Review

The issue of science communications came to the fore once again in the aftermath of Program Review, when EC began taking steps to deal with its impact. The deep budgetary and human resource cuts focused senior management attention on the need to ensure that both the federal government and the public understood the value of the Department's S&T and therefore the return on investments in that activity. The consensus was that there was little awareness or interest in the federal government's S&T activities.

In January 1995, the Department engaged a communications consulting company to develop a strategy for more effectively communicating its S&T. Delivered in March, the strategic plan set out many ideas for increasing the level of recognition and credit accorded to EC S&T.[7] It recommended that the Department build on its foundation of scientific credibility by investing in expanded visibility and a higher volume of communications. In doing so, the report urged the Department to:

> Humanize and personify EC science by empowering, encouraging and equipping department scientists to communicate effectively, consistently and with conviction as they are the sources of sound science behind the programs and initiatives addressing the priority environmental issues.

The Department did not have the resources needed to fully implement the consultant's long list of suggestions. The Communications directorate developed an action plan with four items: a series of short television vignettes, an insert in a science and technology magazine

---

[7] *Effective Communications of Science and Technology: Strategy and Programming for Environment Canada*, prepared by Strategic Reimer Reasons. EC S&T Management Committee, box 2

distributed with newspapers, the use of existing EC promotional print (Enviro-Tipsheet) and radio (Planet Update) media vehicles for stories about S&T leading to solutions, and an EC S&T homepage on the internet. All of these activities featured Departmental scientists, as the strategic plan had advised.

The initiative that attracted the most attention was the television vignettes. The concept was fleshed out in 1996. A series of 12 videos, each about six minutes long, was prepared in partnership with the Discovery Channel. Called *Earth Tones*, it focused on the work of EC's scientists on air, water, wildlife and green technology issues. The first episode, on pesticides, was broadcast during @*discovery.ca* on January 30, 1997. Soon several other science-based departments were interested in participating. This led to the formation in early 1998 of a communications working group under the *Memorandum of Understanding between the Four Natural Resource Departments on Science and Technology for Sustainable Development*.[8] Chaired by EC, the interdepartmental group focused on producing further vignettes reflecting the science for sustainable development performed in the participating departments. It also was engaged in preparing tipsheets, creating a website and sponsoring a workshop at the annual meeting of the Canadian Science Writers' Association.

In addition to these activities, EC Communications worked with the S&T Management Committee to develop two short periodicals that were launched in the fall of 1997. *Science and the Environment Bulletin* shared the results of EC's S&T with the public. The other, *Science and the Environment Issues*, presented the science behind the environmental issues facing Canadians.

When the EC R&D Advisory Board was established, it too was soon drawn into policy discussions about the Department's approach to S&T communications. Among the Board's first working groups was one on communications, chaired by Peter Calamai, a distinguished science journalist. At its first meeting, in January 1998, the group decided to develop a science communications framework for the Department. As background, the group undertook to prepare some case studies on past departmental S&T communications approaches to environmental issues – acid rain and ozone depletion – and to analyze them for lessons learned.[9]

---

[8] A discussion paper, *Federal Science & Technology Awareness*, prepared for NRCan in March 1996 had recommended a joint approach to communications on the part of the MOU departments. See chapter 7 for more on the MOU.

[9] *Communicating Science at Environment Canada: A Brief Review of Lessons Learned from Communications on Acid Rain and the Depletion of the Stratospheric Ozone Layer*. Science Policy Branch working paper #3, February 1999

The Board accepted the framework at its October 1999 meeting. It outlined key principles, management strategies and approaches to guide EC's communication of science.[10] The working group also made five recommendations. EC was advised to adopt the framework's principles and its lessons learned, to make more effective use of the Internet, to provide training in communications for its scientists, and to expand its science communications effort.

Calamai and Lydia Dotto, another Board member and well-known science writer, were also involved in an EC pilot workshop on training scientists in communications. The genesis of the workshop is illustrative of the tensions that sometimes arose between Communications and EC's S&T community. Frustrated by what they perceived as a lack of action beyond the development of plans by Communications, several EC Regional Science leads serving on the S&T Management Committee proposed that they would develop and pilot a practical skills-development workshop for scientists.[11]

The workshop took place from November 30 to December 2, 1998. Looking at communications broadly, it included media training, participation and mentorship by David Phillips and Rob Butler – two EC scientists experienced in communications – and presentations and panel discussions from seasoned pollsters, journalists, TV producers and magazine editors, as well as sessions on voice training, communication in communities, and the importance of considering cultural contexts in S&T communication. Following up on the workshop, a revised three-day curriculum for future training sessions was prepared.[12]

The pilot workshop had been a response to the strategy of using scientists to communicate the value of EC's S&T. It was recognized that most of them were not born communicators and thus required some training. There were several challenges, however, in offering it. Resources were tight; the pilot had been financed through a special Learning Fund set up by the Department to encourage innovation. EC had no training division. And the demand within EC was not big enough to warrant a regular offering. The workshop would not be offered again. However, it did influence the development of another media training course, one that would be offered interdepartmentally.

---

[10] *Science Communications Framework for Environment Canada*, working group on science communications. Environment Canada S&T Advisory Board report no. 2, March 2000
[11] The group was led by Alex Bielak. My thanks to him for this information.
[12] *A Report on an Environment Canada Communications Training Workshop for Scientists and A Science Communications Curriculum for the Future*, by Alex Bielak and Bill Gummer, 1999

Making the Case for Federal S&T

An extraordinary number of negative stories about federal S&T appeared in the press between June and October of 1997 (Figure 10.1). It began on Saturday, June 21 with a story entitled "Bureaucrats sabotage scientists" on the front page of the *Ottawa Citizen*. Referring to a recent article, "Is scientific inquiry incompatible with government information control?" that had appeared in a scientific journal, the *Citizen* reported the claim that DFO had intimidated its own scientists and ignored outside research.[13] It went on to say that "DFO had twisted scientific research to support government statements and released biased, even false information to the public." Over the rest of the summer, the stories expanded to include Health Canada and several other science-based departments.

**Figure 10.1 Some newspaper headlines on federal S&T
from the summer of 1997**

These departments and several others were concerned about the effects of the stories, which were mostly about abuses in the use of science in decision-making. They posed a major challenge to the

---

[13] The article, by Jeffrey Hutchings, Carl Walters and Richard Haedrich, appeared in the *Canadian Journal of Fisheries and Aquatic Sciences* (vol 54, 1997).

government's credibility by undermining claims that decision-making took full account of the best science available. Led by Marc Denis Everell, an ADM and Chief Scientist at Natural Resources Canada, ADMs from the other large science-performing departments – EC, Agriculture and Agri-Food Canada, the Department of Fisheries and Oceans and Health Canada – began to meet at the end of August to discuss actions they could take together to counter the picture being painted of federal S&T. The ad hoc Committee of ADMs on Science in Government, as the group called itself, pursued three initiatives. They sponsored a paper articulating the role of federal science.[14] They developed a code of best practices on the conduct, management and use of government science.[15] And they produced an umbrella communications strategy to emphasize to Canadians the positive value of their S&T efforts. The Committee also secured the agreement of the Clerk of the Privy Council (Jocelyne Bourgon) to give a speech on federal science. She did this at the Federal S&T Managers Forum in December 1998, stressing the importance of communications.

> If there is one area of management that requires our special attention now, it is communications. We need a more concerted and systematic effort to better explain why federal science and technology is important to Canada and to Canadians. We need to take our record of success and communicate it more effectively. We have an exciting story to tell — and it is critical that we tell it. It is our responsibility to explain the importance of science to ministers, to elected officials and to Canadians.[16]

Although EC had not been much implicated in the negative press, the Department was very involved in the Committee's work. EC (through Karen Brown, ADM of the Environmental Conservation Service) led the activity on the communications strategy, probably because of the Department's experience in S&T communications. A workshop was organized in February 1998 involving the ADMs as well as their DGs of communications. Peter Calamai also participated, writing up a short report on what he heard and what issues needed to be resolved.[17]

---

[14] The paper by John de la Mothe, "Government Science and the Public Interest" was published in *Risky Business: Canada's changing science-based policy and regulatory regime*, edited by G. Bruce Doern and Ted Reed (University of Toronto Press, 2000): 31-48.
[15] This activity resulted in a 1999 Health Canada report, *Values and Best Practices: A proposed code for the conduct, management and use of science in the government of Canada*. Science Policy Division files
[16] *Challenges for the Science and Technology Community*, December 1 1998
[17] *Toward a Shared Communications Strategy for Federal S&T*, Science Policy Division files

Following that session, an S&T policy consultant was hired to prepare a detailed action plan. This was ready by June and consisted of 28 suggested projects, including such ideas as media training for government scientists, a magazine on science in government, and field trips to departmental S&T facilities for Treasury Board and Finance analysts.[18]

The consultant's report was given to the communications DGs for their review. They, in turn, involved the interdepartmental communications working group that had been established earlier that year under the MOU on S&T for sustainable development. The result was a five-year action plan funded by the five MOU departments, who were also the members of the ad hoc Committee. In order to keep the group in close touch with evolving federal S&T issues, some science policy advisors were added. In addition, a few months later in January 1999, the small communications effort under another ADMs committee – the TBS S&T Human Resources Framework initiative – was folded into the interdepartmental group, which became known as the 5NR communications working group.[19]

The 5NR group was a collaborative effort aimed at increasing awareness of federal S&T activities among Canadians and the value of those activities in the formation of policy and in other government functions among decision makers inside and outside the federal government. It undertook a number of very successful initiatives, aided by annual financial contributions from the five departments totalling $1.15 million over five years.[20]   One of these was sustaining and extending the partnership with the Discovery Channel on the *Earth Tones* vignettes. These continued to be produced for several years. That effort was then replaced by a one-hour documentary, *First Scientists*, which aired in 2003 and looked at the links between western science and traditional knowledge using examples drawn from departmental activities. There were also several communications spin-offs from the vignettes including a booklet and poster to accompany them, use of the vignettes at departmental exhibits, and production of teaching guides for the videos for grades 7-12 environmental programs.[21]

---

[18] *"Your Resource for the Future": 4NR+H Communication Action Plan*, by Ron Freedman, The Impact Group, June 1998. Science Policy Division files

[19] For more on this ADMs' committee, see chapter 6.

[20] Further details on the communications working group can be found in the annual reports of the MOU. See also the *Review of 5NR Communications Initiatives*, DFO Review Directorate, May 2001, and the report, *The 5NR Communications Working Group: The First Five Years,* June 16 2003.

[21] The booklet, *Earth Tones ... The Book: Federal Science for Sustainable Development*, 2000, won the APEX 2001 Award for Publication Excellence.

A second major initiative was the Science Awards for Leaders in Sustainable Development. This program, which ran between 2000 and 2003, was a competition that recognized scientists in the five MOU departments for their contributions to sustainable development and awarded scholarship supplements to graduate students working in related areas. The third significant effort was a science communications and media training course. Building on EC's experience, the working group ran a number of pilots to develop a course in both English and French. The group then contracted with some communications consultants to make the course available to departments on a fee basis. By 2003 several hundred federal scientists had taken the training. The course, or a variation of it, is still offered.[22]

The action plan guiding the activities of the 5NR communications working group expired in March 2003. A month earlier, a decision had been made not to renew the MOU on S&T for sustainable development, under which the group had operated. While working interdepartmentally on science communications was still valued, the driving force that had led to the formation of the 5NR communications group was no longer there. Science ADMs were more concerned with finding ways to work collaboratively rather than on getting their story out. There was also some fatigue on the part of the MOU departments; the group had required considerable time, effort and financing.[23] And some new S&T communications efforts had arisen. For example, a private sector magazine focused on federal S&T, *Research Horizons*, was first issued in December 2001 and ran until the winter of 2004. And the Federal S&T Cluster (science.gc.ca), which emerged out of the Government On-Line exercise and whose goal was to be a window into online federal S&T, was launched in March 2004. There was no clear path forward for the communications working group.

Communications DGs provided $100,000 for another year to allow the group to develop a strategy for its future. It decided to expand its membership and to ally itself with the ADMs S&T Integration Board.[24] This was agreed to and the one person employed full-time on the 5NR working group was transferred to the Board's secretariat. Due to the broad membership of the Integration Board, the focus of the communications work moved beyond S&T for sustainable development.

---

[22] The course is offered by Rutherford McKay Associates, a communications firm based in Ottawa.

[23] *Next Steps for 5NR Communications: Report on Consultations*, prepared by Ravine Communications Inc., January 2003

[24] *Focus on the Future: Communications Strategy for a Government of Canada S&T Communications Working Group*, presentation to the Integration Board, March 9 2004. Science Policy Division files

Receiving much less funding under the Board than under the MOU, there were also fewer communications activities and these were concentrated on the Board's work. Occasionally, though, a project was undertaken which aimed at promoting the value of federal S&T. In October 2009, for example, an S&T Day was organized on Parliament Hill showcasing the contributions of departmental S&T.

Rejuvenating EC S&T Communications

At the same time as the 5NR communications group was ebbing, EC's Communications directorate was taking steps to reinvigorate its efforts in the area of S&T. A recent Council of S&T Advisors report on science communications in the federal government had recommended that departments develop comprehensive S&T communications strategies and invest in that activity.[25] Referring to the report, the directorate drafted a new strategy in 2003 in consultation with the S&T Management Committee.[26] The strategy's goal, however, was quite conventional – "to raise awareness of the role of Environment Canada's S&T in the formation of policy, regulations and solutions to environmental challenges; showing stakeholders the importance and value of EC science, and ensuring Canadians understand Government of Canada science is part of its service to Canadians." Fulfilling it meant overcoming a number of challenges. The Strategy noted a lack of time on the part of both communications staff and scientists. Few resources were available. And the Department's efforts in S&T communications, while numerous, were uncoordinated and tended to be reactive. To respond to this, the Strategy called for several organizational changes: an S&T communications advisor and working group, a speakers bureau (to identify subject experts and do media training), and a departmental coordination group. The Strategy began to be implemented with three new positions being staffed. But the demands of other communications activities and a large turnover in communications employees led to its being dropped.

While Communications was struggling, the Science Policy Branch acted on the suggestion of the Senior ADM (Robert Slater) and produced a booklet explaining, for a non-specialist audience, why S&T was carried

---

[25] *Science Communications and Opportunities for Public Engagement (SCOPE)*. Report of the Council of Science and Technology Advisors, April 2003

[26] *The Source for Environmental Science and Technology: Communicating Environment Canada's Science and Technology*, fall 2003. Science Policy Division files

out in the Department and how it was managed.[27]   The ideas in this publication were used in many other Departmental communications efforts. Similarly the National Water Research Institute formed a dedicated Science Liaison Branch in 2001. Its mission was to raise the profile of the Institute and to help link science to policy.

A fresh start was taken by Communications in 2008 to build a sustainable S&T communications capacity. Noting that there might be "some degree of skepticism" due to its having been promised before but not delivered, an S&T communications framework was drafted.[28] It was similar to the 2003 strategy including a focus on improved coordination of the various communications efforts in the Department. What was new at the time, as the framework pointed out, was the existence of a Departmental S&T Branch and a strategic Science Plan.

When the S&T Branch was formed in the fall of 2005, a new S&T Liaison Division was also created.[29]   Built on the nucleus of the small group at the National Water Research Institute, S&T Liaison was intended to disseminate the Branch's work to various audiences and to better link it to their needs. The Division undertook a number of communications activities, including improving the Department's S&T website, developing an S&T publishing policy, creating a directory of EC S&T expertise, and producing a series of stories on how EC's S&T had led to tangible benefits for Canadians.[30] The obligation felt by the S&T Branch to communicate S&T beyond the scientific community was also manifested in the Department's 2007 Science Plan. It recognized that transmitting the Department's S&T knowledge to others was an important role of government, and it made commitments to promote more effective communication between scientists and decision makers, to generate a science publications policy and process, and to contribute to improved access to scientific information as well as its management.[31]

Both Communications and the Departmental S&T community had a stake in S&T communications. Their relationship was generally positive and productive. However, as noted earlier, it did sometimes have its challenges usually due to a lack of appreciation for each other's cultural differences. For their part, EC scientists needed to remember their responsibilities as federal employees in interactions with the media. And

---

[27] *Science and Technology: The Foundation for Policy, Regulation and Service*, EC, 2004
[28] *A Communications Framework for Science & Technology in Environment Canada*, presentation to S&T Branch Executive meeting, January 12 2009. Science Policy Division files
[29] This group was led by Alex Bielak who had a longstanding interest in science communications. He had been one of the main organizers of the 1998 EC workshop on training scientists in communications.
[30] The series is titled *Science and Technology into Action to Benefit Canadians.*
[31] *Environment Canada's Science Plan: A Strategy for Environment Canada's Science*, 2007

communications staff did not always factor into their policies the special status of scientific knowledge in society or the general public-interest responsibility of government employees. One example of the friction was the development of a departmental publishing policy in 2006. Making no distinction among types of publications, the first drafts of the policy would have required review by Communications of scientific articles being submitted to peer-reviewed journals. The S&T Branch, after considerable effort, managed to persuade Communications of the impropriety of this approach (let alone its impracticality) and developed a separate standard for S&T publications as part of the Departmental policy.

A second example relates to scientific openness. When EC had been created there was some suggestion that the "objective detachment" of a scientist in the Department might "be hindered by the political necessities of his paymasters."[32] However, the Department had a longstanding tradition of openness in encouraging scientists to deal with the media. As this chapter has shown, many of EC's most successful S&T communications activities were built around its scientists in order to take advantage of their credibility with the public. In February 2008, the practice of encouraging direct interaction between its scientists and the media ended. The new Departmental media relations policy required all staff not to speak with reporters, referring them instead to a media relations officer.

A final example is the publication of scientific material in both official languages. The tension here was created by an interpretation of the Official Languages Act that meant that all scientific publications by government employees would need to appear simultaneously in both official languages. The challenges in doing this were the relatively high expense, the provision of quality assurance, and timely publication, especially in face of the very low demand for translated material. The issue had come up in the 1970s and early 80s, when the mode of publication was printing. A balance between the Act and this form of specialized publication was reached then.[33] However, the advent of government-on-line upset the balance. In 2001, EC had to pull documents from its website because they appeared in only one official language. In September the departmental management committee asked

---

[32] Ralph O. Brinkhurst & Donald A. Chant, *This Good, Good Earth: Our Fight for Survival* (Toronto: Macmillan, 1971). A similar sentiment was expressed in an editorial in the magazine *Science Forum*, issue 23, vol 4, no 5 (October 1971). The latter was written by K. A. Kershaw who, like Brinkhurst and Chant, was a university professor.

[33] The issue had been discussed at meetings of the departmental management committee on December 12 1974, March 22 1979 and December 1982. LAC Acc. 1991-92/017, boxes 11, 29 & 40

the S&T Management Committee to look at the tension between open sharing of scientific articles and reports, and the desirability of making that information available in both French and English. The Committee produced a report, *S&T Communications and Official Languages: Challenges and Options for the EC Community*, which was discussed by the departmental management committee on April 9, 2002.[34]   Based on this, the Committee then produced a set of guidelines, *Official Languages and S&T Communications: Interim Guidelines*, which were presented to the departmental management committee in October.[35]   The guidelines would have standardized scientific translation practice in the Department and would have seen more material being translated. However, they would not have resulted in departmental practice being fully compliant with the Act. As a result, implementation of the guidelines did not go forward and the issue was studiously ignored. As these three examples illustrate, despite the synergies of working together to communicate government science, tensions can arise due to conflicting values and objectives.

---

[34] Science Policy Division files
[35] Ibid.

11

Linking Science and Policy

The value of using science in the development of policy and in decision-making has been widely held in EC throughout its history. Statements about the role of the Department's science always included the development of environmental policy. The ability to link science to policy was said to be the single most important competence in the Department.[1] Indeed, the role of advising on the implications of science for policy has a strong claim to be the defining norm of science in government in contrast with science in academe or in industry.

Work at the science-policy interface was a pervasive, ongoing and intrinsic part of EC's activities. It went on in various parts and at different levels of the Department, and was generally overseen by EC's management systems. For some highly controversial issues such as climate change and endangered species, separate formalized science advisory processes were put in place. Most of the time, the work went on without much notice as part of the Department's regular business.

Despite the importance of linking science to policy for EC and the numerous and widespread activities devoted to it, the topic itself was rarely the subject of policy work at the departmental level. The main exception was from 1997 to 2002, brought about by a controversy centered in two other science-based departments. Nonetheless, EC senior managers were active in dealing with the challenges. The proper use of science for policy was important to the work of the Department, and to maintaining public trust.

EC work on the science-policy interface was able to draw on the Department's rich and extensive experience with the interaction of science and environmental policy. That the work was motivated by the government's attempts to deal with miscues at the interface, rather than by the importance of the topic to EC, reflects a lack of ongoing effort by the Department to learn from and improve upon its own experience. EC management tended to be preoccupied with short-term activities aimed at environmental policy outcomes, to the neglect of the Department's overall management.

---

[1] Comment by Robert Slater, as noted in *Science and Technology Advice: From Framework to Implementation*, presentation to the EC Management Board, May 13 2002. Science Policy Division files

Engaging Scientists

The importance of EC scientists and their scientific knowledge to policy-making was clear from the beginning of the Department. Meyboom's 1972 discussion paper, proposing a departmental science policy, devoted one of its three chapters to the role of scientists in government.[2] He called for stronger links between scientists and policy makers, because he saw the Department's effectiveness as dependent on the two groups having a mutual understanding of their respective roles and values. His view about the importance of scientists in providing policy advice appears to have been widely shared in EC. Comments on his paper, some of which were circulated with it, generally agreed on this point. In addition, a report published the following year on research in the Inland Waters Directorate concluded that the research function at EC "exists to provide information required to develop and implement new policies and to implement existing policies more effectively."[3] It also called for better representation of research scientists in the "policy making and implementing functions."

Perhaps because there was a general consensus on the need to link science and policy, and since the links were largely managed by the Services, it was not the subject of much departmental policy work in EC's first two decades. Occasionally, there was a call for greater attention to the topic. For example, the Minister's Canadian Environmental Advisory Council recommended in 1977 that "steps be taken to develop environmental research specifically aimed at formulating policies for environmental protection and quality" and not just for the management of renewable resources.[4]   And the 1984 *Policy Respecting Science and Technology* committed the Department to developing procedures to involve scientific personnel in decision-making.

Although there was little policy work on the subject, linking science and policy was a core function in the Department. It could easily come to the forefront of issues given the right circumstances. This happened in the early 1990s with the decision to conduct departmental science assessments. Concerned with better integrating the Department, the DM asked for a strategy for EC's R&D in 1991.[5] Work on this led to, among other things, the Department's first science forum. Held the following year, it was designed to help involve scientific researchers in setting

[2] *Science in a Changing Environment – proposals for a departmental science policy*, January 1972
[3] J. A. Gilliland & A. D. Stanley, *Research in the Inland Waters Directorate*, [1973]
[4] Philippe Garigue, *Federal Research and Policy-Making in Environmental Affairs*, February 1977. LAC Acc. 1992-93/011, box 55
[5] See chapters 1 & 2.

strategic directions for R&D. The forum saw considerable discussion about whether and how science was used in Departmental decision-making. The result was a recommendation that science assessments be used as a standard departmental practice in order to ensure the input of science into policy.[6]

Following the forum, the Office of the Science Advisor established a departmental group to more clearly define science assessments and to develop guidelines for their conduct.[7] The group also suggested a number of potential topics for assessment. The Departmental Management Committee decided, in February 1993, to use science assessments as a formal Departmental tool to link research and policy, and chose biodiversity as the subject of the first assessment.[8] That topic was selected because Canada had recently ratified the Convention on Biological Diversity and EC was responsible for the preparation of a Canadian Biodiversity Strategy. It was thought that a science assessment would help EC, as well as other departments, in formulating action plans to implement the Strategy.

Biodiversity in Canada: A Science Assessment for Environment Canada was published in 1994. It was a very thorough study consisting of 13 chapters totalling 245 pages, and was the product of a large team of authors many of whom were drawn from outside the Department. Although there were plans to do further assessments – there was a list of 18 possible candidates – *Biodiversity in Canada* turned out to be the last science assessment of its type.

Science assessments had not been a new concept in EC. Many had been carried out before and many would be carried out in various units within the Department after the biodiversity science assessment. Conducting a departmental science assessment was expensive both in terms of money and the time of scientists. Moreover, the acceptance by senior management of this tool had taken place in the context of a preoccupation with forging a more unified Department. By 1994, the advent of Program Review had shifted the focus to other challenges. Science assessments would not be used again at the departmental level as a vehicle for assessing the implications of science for policy issues.[9]

---

[6] *Synthesis Report. Science Forum I.* Office of the Science Advisor, December 1992
[7] *Environment Canada Science Assessments*, March 1993. Science Policy Division files
[8] Meeting, April 6 1993. EC S&T Management Committee, box 1
[9] The Government of Canada supported the creation of the Canadian Academies of Science (later renamed the Council of Canadian Academies) in 2005 in order to conduct independent science assessments. Currently, this organization is working on two assessments for EC.

Science Advice in Government

Before the late 1990s, EC's activity to more effectively link science and policy had been prompted mostly by a desire to better engage its scientists. This changed in the summer of 1997 when a major controversy erupted about the use of federal government science in policy development.[10]   A large number of stories appeared in the media about abuses in the use of science in decision-making. The main targets were the departments of Fisheries and Oceans and of Health, but EC was not completely unscathed. Because of the importance it attached to strong links between science and policy, the Department took the challenge posed by the controversy very seriously. The negative press had largely disappeared by the fall, but did occasionally reappear. Government activities to counter it would last for another five years. EC would be a major player in efforts to deal with the controversy and to strengthen science-policy linkages.

EC collaborated with other major science-performing departments on a number of actions to restore public confidence in the government's use of science. It also engaged its newly created, external R&D Advisory Board on the issue. At its second meeting, in October 1997, the Board decided to create a working group on science advice in government. The group's mandate was to provide advice on mechanisms for effective, transparent linkages between science and policy, and on a process for developing a statement of ethics for federal science.

With regards to its first charge, the working group reviewed EC's experience with various types of science assessments. It recommended that EC evaluate its science assessment process as a means of documenting the link between science and policy. A report on the evaluation, focusing mostly on the experience of the Atmospheric Environment Service, was completed by November 1998.[11]   In the area of ethics, the working group monitored the interdepartmental work on a code for the conduct, management and use of science.[12] The group also proposed that EC name a science ombudsman to serve as a sounding board for scientists who felt their views were misrepresented or ignored by policy makers. The DM (Ian Glen) did not think that the situation within the Department at that time warranted this appointment. He believed that existing departmental policies for dealing with conflicts – such as the grievance process, disciplinary policy, harassment in the

---

[10] For further details see chapter 10.
[11] Elizabeth Bush, *Science Assessment: A Report on Science-Policy Linkages in the Atmospheric Environment Service*, November 15 1998
[12] See next section.

workplace, and the mediation program – were sufficient. Nonetheless, he asked the Department to confirm that the functions of an ombudsman were being fulfilled by those policies.[13]

The controversy had also provoked the formation of the federal Council of S&T Advisors, which the government had committed to in its 1996 S&T strategy. Upon establishment in April 1998, the Council was asked by the government to develop a set of principles for the effective use of science advice in government decision making. The Council considered efforts in other countries to better link science and policy, especially work done the previous year in the UK on principles to guide science advice.[14] It also looked at practices in the federal government.[15] The Council released its report, *Science Advice for Government Effectiveness (SAGE)*, in June 1999. It included six principles – early identification, inclusiveness, sound science and science advice, uncertainty and risk, openness, and review – each supported by guidelines, and some advice on implementation.

With the release of *SAGE*, an interdepartmental group began work on a government response, including a series of consultations. EC's R&D Advisory Board was well informed about the Council's work, since two of its members were also on the Council.[16] The Board decided in July to review EC's current efforts related to the six principles and to make recommendations about how the Department might further implement them. The Board's report was presented to EC in October.[17] The Department agreed with all of its recommendations.[18]

While waiting for the government's response to *SAGE*, EC also conducted a review of its science advice practices.[19] The study was done in three parts. The first was an inventory of the major practices. Sixty-one measures were listed, organized according to the six principles in *SAGE*. The second part analyzed the measures and found that they generally met

---

[13] *Overview of Environment Canada's Problem-Solving Approaches and Science Initiatives*, September 1999. Science Policy Division files

[14] Sir Robert May, *The Use of Scientific Advice in Policy Making*, UK Office of S&T, March 1997

[15] *Scientific Advice in Government Decision-Making: The Canadian Experience*. A report in support of the work of the Council of S&T Advisors, March 1999. This includes a profile of the work being done at the time in EC.

[16] The Council membership was drawn from the external advisory bodies to S&T-based departments and agencies.

[17] *Science Advice for Government Effectiveness: Recommendations for Implementing the SAGE Principles at Environment Canada*. S&T Advisory Board report #3, November 1999

[18] *Response to the Recommendations of the S&T Advisory Board on Implementing Science Advice Principles at Environment Canada*, March 2000

[19] *Science Advice in Environment Canada*, prepared by Alex Chisholm. Science Policy Branch working paper #11, 2000

the requirements of the principles and guidelines. It also proposed some options for improving EC's performance. Part three looked at the role of EC's business lines in linking science and policy. EC used business lines to foster a more department-wide approach to managing its priorities. The study found some evidence that the business lines were also serving to increase communications between the science and policy communities in EC. The business lines had the potential to foster "more focused science and better informed policy."

The review had focused on EC's existing science advice practices. It did not look at areas where there were limited or no measures. This gap was examined early in 2001. Some issue areas were identified where there was limited R&D or where science-policy linkages needed strengthening. The state of these links, however, was constantly changing. When the study was repeated a year later, most of the issues had been or were being addressed and new ones were surfacing.[20]

The prominence given to the relationship between government science and policy in the late 1990s also led to interest in the issue by other policy research groups. This created other fora in which EC was involved. The independent, not for profit Public Policy Forum commenced a series of activities on science in government in the spring of 1998, of which several were on science advice. It produced a discussion paper on the role and responsibilities of the scientist in public policy, and held a high-level roundtable for senior bureaucrats and CEOs on links between science and policy.[21] As well as participating in the roundtable, EC provided advice on its agenda and some financial support towards the event. Another example was the federal Policy Research Secretariat, which sponsored a workshop on science and scientists in policy development in May 1999. Several EC scientists, science managers and policy advisors participated.

Framework for S&T Advice

The government released its response to *SAGE* in May of 2000, in the form of *A Framework for Science and Technology Advice*.[22] The document was largely a replication of *SAGE*. The principles were essentially the same.

---

[20] *Science Policy Linkage in Environment Canada*, PowerPoint presentation, June 2002. Science Policy Division files

[21] *Blood, Fish and Tears: A Round Table Discussion on the Credibility and Acceptability of Science Advice for Decision-Making. Summary of Discussions*, Public Policy Forum, November 1998

[22] *A Framework for Science and Technology Advice: Principles and Guidelines for the Effective Use of Science and Technology Advice in Government Decision Making*, Industry Canada, 2000

But some adjustments had been made to the guidelines, and the implementation measures had been fleshed out, reflecting the consultations that had taken place over the course of the year since *SAGE*. In addition, the government declared that departments had about three years, until the end of March 2003, to implement the *Framework*. All departments were required to adopt the implementation measures, but they were allowed the flexibility to deal with the principles and guidelines in ways appropriate to their mandates and situations.

With all the work that EC had already done to assess its science-policy practices against *SAGE*, it was very well prepared to develop an implementation plan for the *Framework*. The plan was presented to the Departmental management committee in December.[23] It proposed two sets of activities to be carried out over three years. The first consisted of actions the Department could take to carry out the mandatory implementation measures set out in the *Framework*. The second was a list of actions that might also be pursued on the principles and guidelines. It was composed of suggestions, based on EC's earlier studies, to address issue areas where there was limited R&D (e.g., ecosystem effects of genetically modified organisms) or to strengthen science advice processes (e.g., integrating health and environmental issues in policy making). The Departmental management committee approved the implementation plan. Business lines were assigned responsibility for carrying out the plan, and the S&T Executive Committee tasked with overseeing their progress.

Some of the actions in the implementation plan were identified as items that would best be pursued jointly with other departments. The interdepartmental ADMs Committee on S&T shared this view. It created a subcommittee on S&T advice, which commenced work in the spring of 2001. Probably in acknowledgement of the extensive work the Department had already done on the subject, the committee's chair was the EC ADM (Karen Brown), who was the Departmental lead on science policy, and EC's Science Policy Branch provided the secretariat.

Over the next year and a half, the subcommittee developed four tools to assist departments in implementing the Framework. One was the organization of a best-practices workshop. It was designed to share, through case studies, departments' experience in linking science and policy and how that aligned with the *Framework*. The workshop took place on October 17, 2001 and attracted about 100 participants from over 15 departments and agencies.[24]

---

[23] *An Implementation Plan for Environment Canada: Framework for Science and Technology Advice*, December 11 2000. Science Policy Division files
[24] For a list of the case studies, and for further information on the subcommittee's activities, see "Implementing the Framework for Science and Technology Advice,"

The second tool was a training course on S&T advice in policy. EC and NRCan contracted with Bruce Doern of Carleton University to design a two-day course and prepare a training manual. The course was piloted twice, in October and in November, and included a mix of theoretical instruction and case studies. The pilots helped to refine the course, which was then made available for others to use.[25]

The third tool developed by the subcommittee was a "Science Advice Checklist for the Preparation of Memoranda to Cabinet and for Regulatory Impact Assessment Statements."[26] It was prepared and tested by a working group led by the Privy Council Office.[27] The checklist could be used as a planning tool, to help ensure that science advice was properly used. However, its intended purpose was accountability. It was a way of demonstrating to senior departmental officials and ministers that science advice had been properly considered in policy and regulatory formulation. Yet, there was no requirement that a completed checklist be submitted along with the documents to Cabinet. All that was necessary was an acknowledgement that the *Framework* had been adhered to.

Some of the lessons learned and material developed for the first three initiatives were incorporated into the final one, a guide on implementing the *Framework* for science and policy managers.[28] The initial idea had been to prepare an evaluation guide for assessing government conformity to the *Framework* and its impact. But the subcommittee decided to focus on assisting managers. It developed a self-assessment worksheet to help them determine how their science advice practices rated against the principles and guidelines of the *Framework*. The guide also included other useful information for managers, such as examples of good practices and a list of frequently asked questions.

EC had played a major role in the work of the interdepartmental subcommittee. Many of the latter's products were incorporated into the Department's activities to fulfill its *Framework* implementation plan. For example, an amended version of the checklist was produced to reflect EC's broader needs (see Figure 11.1). And an online resource, *Science*

chapter 3 in *Science and Technology Advice: A Framework to Build On. A Report on Federal Science and Technology*, Industry Canada, 2002.
[25] *Science and Technology Advice in Policy. A Pilot Course Prepared for Natural Resources Canada and Environment Canada*, November 2001. Science Policy Division files
[26] A copy of the checklist, as well as an example of a completed version, can be found in the guide referenced in footnote 28, below.
[27] Starting in 2003, the checklist and manager's guide were available on the Privy Council's website. At some later point they were no longer available there.
[28] *Implementing the Principles and Guidelines of the Framework for Science and Technology Advice: A Guide for Science and Policy Managers*. Science Policy Branch working paper #17, September 2002

Policy Integration (SPI), was built which consolidated many of the pieces produced by the subcommittee. *SPI's* purpose was to help a broader audience learn about the *Framework* and about the process of linking science and policy. It did this through an interactive learning website that contained background information, the principles, case studies, the self-assessment worksheet, the checklist, and other material. *SPI* was available to all EC employees through the Department's internal website.[29]

In addition to its work on the subcommittee, the Department undertook other activities as a result of its interest in the science-policy interface. It worked with several departments to produce a manager's guide for assessing the impact of science on policy development.[30] It also engaged in a significant S&T values and ethics initiative.

As a result of the controversy in the summer of 1997 – which had been the major factor in sparking *SAGE* and the *Framework* – EC had worked with several other science-performing departments to develop a draft values and ethics document on the conduct, management and use of science in the government.[31] However, EC did not adopt the draft, thinking it too skewed towards the context and experience at Health Canada. The Department decided to develop its own guidance once the *Framework* was released.

Assisted by some funding from the Department's Learning Fund, EC's own effort on S&T values and ethics was launched in 2000. A total of seven workshops, involving over 100 S&T employees, were held across the country between late 2000 and the first few months of 2001. The workshops examined values and ethics issues in EC's science effort.[32] They revealed that employees held themselves and the Department to a very high standard. The workshops also showed that many of the ethical issues with which staff were concerned related to the science-policy-politics interface. The output of the workshops included seven suggestions for the Department's consideration. All of them were addressed in the next few years through, for example, the creation of an orientation website, the naming of a Departmental values and ethics

[29] *SPI* is accessible to government employees on GCpedia.
[30] Bronson Associates, *A Managers' Guide for Assessing the Impact of Science on Policy Development*, 1999. The work was done in collaboration with EC, NRCan, DFO, HC, and AAFC. Science Policy Division files
[31] *The Best Practices Initiative: Best Practices for the Conduct, Management and Use of Science in the Government of Canada*, Health Canada, 1999. This work did influence part of *SAGE*.
[32] *Science in the Public Interest: Values and Ethics in the Management, Use and Conduct of Science at Environment Canada*. Science Policy Branch working paper #15, March 2001

Issue Identification	Inclusiveness	Sound Science/Science Advice	Uncertainty and Risk	Transparency and Openness	Review
1. What is the background and magnitude of this issue? How was the issue identified? What are the implications? How and to whom was the issue communicated for action? 2. Are S&T considerations important for the development of policy options? If yes, what role do they play in the development of policy options? 3. Will this issue raise legal, moral or ethical questions which need to be addressed? Outline. 4. What is the degree of public knowledge and understanding of the scientific background to this issue?	1. What are the sources of science and science advice: in-house expertise; external expertise, international; expertise, or a combination of the above? What process was used to provide the advice? Was the advice solicited or unsolicited? Outline advice and specify source. 2. Have the scope and implications of the scientific basis for this issue been explored with related disciplines and departments including social sciences and sources of traditional knowledge? What were the results? 3. What measures were taken to ensure that the advisor(s) chosen matched the nature and breath of judgment required? 4. What measures have been taken to avoid (or manage) potential or real conflicts of interest on behalf of the advisors? 5. Does the advisory process include provisions for minority views or dissenting opinions?	1. What in-house expertise is available to assess and communicate the science and science advice to policy advisors and decision makers? 2. What measures have been taken to ensure the quality, integrity and objectivity of the science advice provided? Has in-house scientific research been subjected to peer review? 3. Have inherent biases been managed? 4. How were science advisors involved in the identification and assessment of policy options? How was their advice reflected in the options presented to decision-makers? 5. How were conflicting scientific views managed in the provision of science advice?	1. What is the nature and degree of the scientific and technological uncertainty and risk of this issue? 2. How were scientific and technological uncertainty dealt with in formulating policy options? 3. How and when were the degree and nature of scientific uncertainty and risk communicated to decision-makers, stakeholders and the public? 4. How was the government's integrated risk management framework applied in addressing this issue? 5. What risk management approach was used in reaching decisions? Outline.	1. How was the need for effective consultation balanced with the need for timeliness in decision-making? In particular, were key interest groups, other government organizations, and international organizations given early notice of significant policy and regulatory initiatives? Outline. 2. Was a representative stakeholder group selected to comment on the development of policy options? How was this group chosen? What were its views and how were they taken into consideration? 3. How were the scientific findings, analyses and policy advice made available to the public and to stakeholders in a timely and ongoing manner? 4. What public consultation was undertaken on the policy options? How have public concerns been taken into consideration?	1. What tools and mechanisms are in place for monitoring, measuring and reporting on the scientific implications of the policy? 2. What are the provisions for a review of the science and decisions (based on a set time period or on significant changes in the science or policy)?

Figure 11.1　EC checklist for science advice, 2002

champion, and the development and adoption of the new Treasury Board Policy on Disclosure of Wrongdoing.

EC was also heavily involved in one other significant activity to bridge science and policy at that time, a roundtable on science and public policy sponsored by the federal Canadian Centre for Management Development.[33] The members of the roundtable believed that a new paradigm was needed, one that integrated science and policy functions around key issues. Their report looked at the issues hindering this "common purpose," proposed four cornerstones for progress (roles within the science and policy communities, organization of work, training and development, and rewards and recognition), and made some suggestions for action on each of these.[34] The results and perspective of this work were incorporated into EC's various efforts in implementing the *Framework*.

The Department's management committee had been receiving progress reports on the implementation plan. By December of 2002, the working group was reporting that EC had successfully tackled all the mandatory measures as well as many of the initiatives the plan had listed as good to do.[35] There were still other items that could be pursued, but the Department was in a position to say that it had adopted the *Framework* and taken many additional steps to strengthen its links between science and policy.

Policy Pull for Science

The attention paid to the issue of science-policy linkages in the wake of the 1997 controversy had mostly run its course by the end of 2002. In EC it was subsumed under a new Departmental initiative on knowledge management, known as "Knowledge in the Service of Canadians." However, science for policy was never in danger of being forgotten, as it was central to the management of the Department. The DM's S&T management review panel, for example, brought up the question of

---

[33] The Centre was a federal organization supporting the continuous training, development and learning needs of public service managers. Its current name is the Canada School of Public Service.

[34] *Creating Common Purpose: The Integration of Science and Policy in Canada's Public Service*, Canadian Centre for Management and Development, 2002

[35] *Federal Framework for Science and Technology Advice. Environment Canada's Response to the Implementation Measures. Status Report*, presentation to the Environment Management Board, December 2002. Science Policy Division files

bridging science and policy in 2004, and recommended the development of tools to facilitate such links.[36]

At the same time as the panel's report, the Science Policy Branch had organized a workshop on the science-policy interface. It was part of an initiative on peer-assist workshops, which in turn was carried out within the Department's work on knowledge management.[37] The workshop led to another in March 2005. Participants felt that the science elements of the science-policy process were relatively well understood, the policy aspects less so. This then became the focus for a December workshop on understanding the role of the policy analyst in the translation of science to policy.[38] Several non-governmental experts explored recent research on the subject along with EC staff.

After the December workshop, selected EC policy analysts were interviewed on their background and their work at the science-policy interface. The survey results were discussed at two subsequent workshops. Several projects were designed to strengthen the capacity of the policy community. However, budget cuts in 2007 prevented most of them from going forward, and delayed the others.

One of the themes that had come out of the workshops and meetings was the role of brokers in translating between the science and policy communities.[39] In recognition of this, EC's 2007 *Science Plan* made a commitment to promote more effective communication of science based on its use in policy development. Two initiatives, in particular, were undertaken: EC Science Alert and EC S&T Expert. The first provided summaries of EC's research results for the management, policy and communications communities, while the second was a directory of Departmental expertise.[40] As well, a small international workshop, "Brokering Knowledge for the Environment," was held in September in partnership with Sarah Michaels, then a professor at the University of Waterloo.

The examination of the policy community in science-policy linkages proved to be short-lived. The emphasis within the S&T Branch on knowledge brokering shifted the focus of policy work away from the policy community and what it needed to do to better interact with the

---

[36] *Science and Technology Management Review Panel Report*, November 2004

[37] S. Pope, M. Friesen and S. Board, *Making Connections that Work*, EC, 2005

[38] *Understanding the Policy Analyst at the Science-Policy Interface*. Science Policy Branch working paper #40, 2005

[39] A. Bielak et al., "From Science Communications to Knowledge Brokering: The Shift from Science Push to Policy Pull," in *Communicating Science in Social Contexts: New Models, New Practices*, eds. D. Cheng et al. (Springer, 2008), 201-226

[40] The two activities were among those that had emerged from discussions on the interview results. For more on policy for S&T communications, see chapter 10.

scientific community. Policy attention moved back to science and to ways of communicating the results of research to a larger audience.

Epilogue

Over the years, I have often been asked during introductions or at social occasions about my occupation. My response, "I work in science policy," has usually elicited puzzled looks and awkward pauses in conversation. The lack of understanding of what constitutes policy work in science, I have come to realize, is completely normal. Like other areas of policy work inside government, its activities are obscure. Regrettably, there are few opportunities for learning about it. Not much has been written on policy for science, and even less that attempts to follow that subject in federal departments over many decades.

This book is my contribution to changing that situation. It reveals, through the means of history, what has been policy work in the area of science in a department of the government of Canada. The individual chapters have highlighted various lessons from Environment Canada's experience. This epilogue pulls them together. It aims at summarizing what that past tells us about doing policy for science.

Its history shows that Environment Canada has been engaged in policy for science since the Department's origin in 1971. The ongoing requirement for science to deal with the growing number and complexity of environmental issues provided the basis for that work. The magnitude of its scientific activities ensured the need for policy for science. The ultimate goal of the policy effort was to ensure that Environment Canada had the scientific capacity it needed and that its scientific activities were conducted efficiently and effectively.

Policy work on Environment Canada's science was episodic, driven by both internal and external factors. Internally, it was carried out in response to specific departmental management needs, the struggles of senior managers to get the most from tight or declining budgets, and the ambitions and frustrations of the Department's scientific personnel. Externally, federal government policies for science, the administrative requirements of central agencies, and evaluations of the government's policies and practices were the main catalysts for the Department's activities in policy for science.

Doing policy for science in Environment Canada was not a routine activity. Policy work in government is pragmatic, like politics it is the art of the possible. It is also essentially a process, consisting of meetings, suasion, and the exercise of authority. Relating that process is just as important as focusing on its products in presenting a history of the Department's policy for science.

Various features of the Department's context restricted the policy response. The focus on environmental issues often resulted in reduced attention to science. The division of scientific activities among its Services challenged the development of department-wide policies.

The policies and practices of the federal government were also significant factors in Environment Canada's policy work on science. They sometimes catalyzed action in the Department. At other times they created complications or barriers. The Department's efforts were often faced with the government's lack of interest in its own scientific capacity, its fixation on the economic contributions of science and technology to the neglect of their other roles, and its weak coordinating mechanisms.

The organization of departmental policy work on science evolved over the years. As the Department slowly became more integrated, responsibility for policy for science moved from its policy group (which came to focus on environmental policy) to its science group. That history shows that science presented the Department with a significant management challenge with which it has struggled over the years. Environment Canada had difficulty in finding a governance system that was able to devote sufficient attention to both its scientific foundations and its environmental mission, to both its means and its ends. Because science was usually regarded simply as a means to an end, its management tended to be overlooked by the Department's administration. Pressing environmental issues largely monopolized senior managers' deliberations.

The policy issues on which the Department worked can be sorted under four headings: general strategy, capacity, collaboration and communications. Policy work on overarching strategies for science usually did not meet with great success. Producing long-term strategies in a government context was always challenging. Sometimes, as in the cases of northern and international S&T, the lack of Department-wide strategies for its environmental activities in those areas created a barrier. And even when a departmental strategy could be formulated, there were impediments to its implementation. One of these was the issue of integrating science strategies into the Department's planning and reporting systems. Another major difficulty was the lack of demand for such science strategies from the federal government.

The need to make the case for its scientific capacity was a primary focus of Environment Canada's policy work. That effort struggled with a number of challenges. Among these were the subordinate position of science in the Department, its fragmentation into Services, and the inherent difficulty in determining the level of capacity that was required. But the biggest challenge came from the federal government. Its struggles with deficits and its inclination to build up S&T capacity in the university and industry sectors, usually at the expense of federal departments, created a major barrier to the effectiveness of efforts in the area of S&T capacity. Policy work was necessary to define criteria for what S&T should be done by the Department, to assess its S&T requirements and to provide arguments for increasing capacity. But it was not sufficient to deliver the needed resources.

Collaboration was another major focus of Environment Canada's effort in policy for science. The interest stemmed from the crosscutting nature of environmental issues and from the increasingly collaborative nature of the scientific enterprise. In addition, tight financial resources put a premium on leveraging and finding synergies in attempting to respond to a growing number of environmental priorities. The Department was very innovative in its efforts. However, the advantages of collaboration were offset by several factors. Budget cuts usually led to a narrow focus on core missions. The latter did not always align well with the priorities, usually economic, of other departments and of the government. And, perhaps most important of all, the ability of the federal government to enable coordination across its departments and agencies was weak.

Policy work on communicating S&T was not as extensive as that in the other three areas. Explaining the role and value of Environment Canada's S&T and improving the use of science in the development of policy and in decision-making were not usually priorities. Despite their importance, the Department was reluctant to pay attention and dedicate resources to them. Activity was usually stimulated by external pressures such as criticism of the management of science, requests for reporting, budgetary cuts and, sometimes, negative publicity. However, the Department's limited policy work here once again proved to be innovative, drawing on its rich and extensive experience with science communications and with the interaction of science and environmental policy. At the same time, the effort faced major challenges in the government's policies dealing with communications and a culture that did not promote openness.

The perseverance of the Department's policy work on science over the years, even with such a challenging context, is a testament to the importance it attached to its science and to the initiative of its managers and policy analysts. Those efforts allowed the Department to deal with serious issues and led to some innovative policies. They helped to enrich thinking and discourse about the Department's science and its management. But they also were often stymied. On the whole, despite a strong engagement in policy for science, Environment Canada was unable to effectively integrate the strategic management of science into the Department's management system. That failure reflected, to a certain extent, the federal government's neglect of science in the public service.

The events covered in this book occurred between the foundation of Environment Canada in 1971 and 2010. The Department has not ceased to exist. Its activities, including policy work on science, go on. Since 2010, efforts have been made to develop a new S&T strategy, to secure funding for climate research, to support the work of the Integration Board, and to establish a policy on science integrity. The lessons of the past tend to be quickly forgotten in government departments. Corporate memory is generally neglected as management attention is usually occupied with the near-term future. The book should help to counteract that memory loss. At the same time, a greater awareness of Environment Canada's policy work on science and the lessons to be learned from its history do not guarantee that new efforts will be easy or successful. However, I trust that the book will be of some assistance to those shaping the future environment for science in the Department as well as in the federal government.

APPENDICES

Appendix 1: Expenditures[1]

EC's total expenditures, 1971-72 to 2010-11

The data shown here are taken from the actual expenditures entries (in millions of dollars) for EC in the annual estimates to Parliament and, in more recent years, from annual departmental performance reports. The chart also presents the equivalent of the expenditures in constant 1972 dollars, revealing a decline over the 40-year period. However, the drop is offset by the fact that the Department was much smaller in 2010-11 than when it was created, with both the fisheries and forestry components having been removed from EC during that time. The Department had about half as many employees in 2010-11 as in 1971-72 (see Appendix 2).

---

[1] My thanks to Matthew Wallace for designing the charts in this appendix.

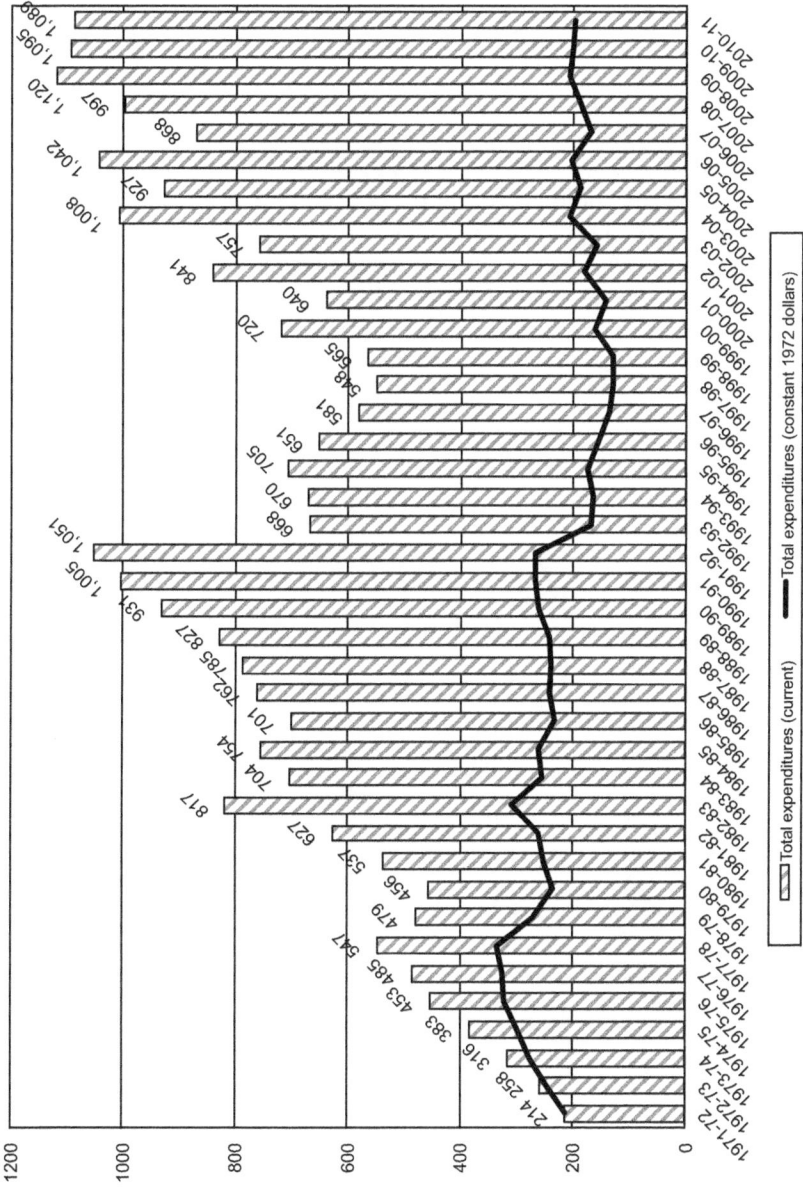

Total expenditures (current) —— Total expenditures (constant 1972 dollars)

1971-72 214
1972-73 258
1973-74 316
1974-75 383
1975-76 453
1976-77 485
1977-78 541
1978-79 479
1979-80 456
1980-81 557
1981-82 627
1982-83 817
1983-84 704
1984-85 754
1985-86 701
1986-87 762 785 827
1987-88
1988-89
1989-90 937 1,005 1,051
1990-91
1991-92
1992-93 668 670 705
1993-94
1994-95
1995-96 581 651
1996-97
1997-98 548 565
1998-99
1999-00 720
2000-01 640
2001-02 841 757
2002-03
2003-04 1,008 1,042 927
2004-05
2005-06
2006-07 1,120 997 868
2007-08
2008-09
2009-10 1,095 1,080
2010-11

EC S&T and R&D expenditures, 1971-72 to 2010-11

The data in this chart are drawn from a number of different reports on federal government S&T expenditures. Prior to 1990, I have relied on reports from the Ministry of State for Science and Technology, prepared by Statistics Canada. For the period after the 1990s, I have used Statistics Canada reports. It should be noted that several different series of reports were issued by the Ministry and by Statistics Canada. These do not always agree on the expenditures in a given year, especially before 1990. In piecing together the data, I have used actual expenditures rather than projections and chosen those most consistently reported. Note that the vertical axis is in millions of dollars.

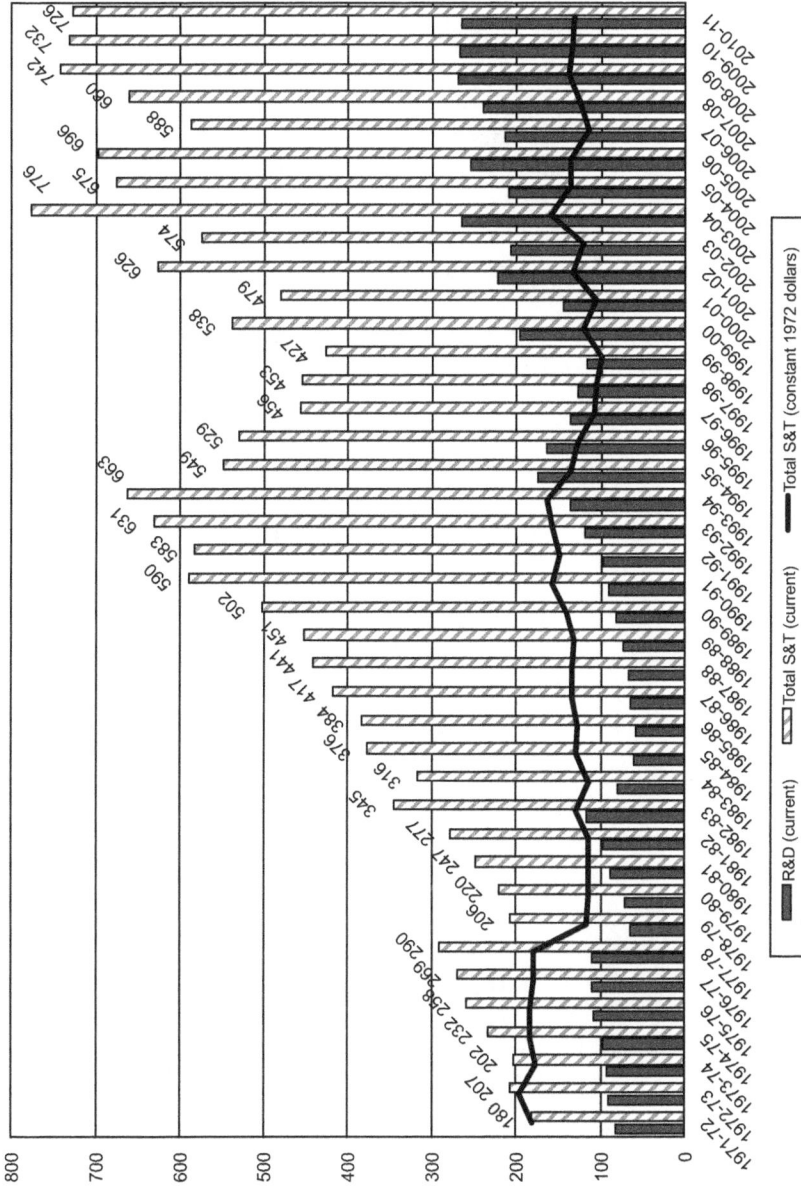

Legend: R&D (current) | Total S&T (current) | Total S&T (constant 1972 dollars)

Y-axis: 0, 100, 200, 300, 400, 500, 600, 700, 800

X-axis (years): 1971-72, 1972-73, 1973-74, 1974-75, 1975-76, 1976-77, 1977-78, 1978-79, 1979-80, 1980-81, 1981-82, 1982-83, 1983-84, 1984-85, 1985-86, 1986-87, 1987-88, 1988-89, 1989-90, 1990-91, 1991-92, 1992-93, 1993-94, 1994-95, 1995-96, 1996-97, 1997-98, 1998-99, 1999-00, 2000-01, 2001-02, 2002-03, 2003-04, 2004-05, 2005-06, 2006-07, 2007-08, 2008-09, 2009-10, 2010-11

Data labels: 180, 207, 202, 232, 232, 238, 269, 290, 206, 220, 241, 277, 345, 316, 376, 384, 417, 441, 451, 502, 590, 583, 631, 663, 549, 529, 456, 453, 427, 538, 479, 626, 574, 675, 776, 696, 588, 680, 742, 726, 732, 726

Proportion of EC Expenditures devoted to S&T, 1971-72 to 2010-11

This chart shows that the majority of EC's expenditures over its first 40 years were devoted to S&T activities. It should be noted that the data used are derived from two different sources. Data on Departmental expenditures are from reports to Parliament, those on S&T expenditures from reporting to Statistics Canada. Although both data sets were assembled by EC's financial services, the latter's accounting systems were organized to provide reports to Parliament. Its calculations of S&T expenditures were always a special effort. Reviews of the S&T data by the Science Policy Branch have confirmed their general accuracy. Still, data on S&T expenditures should be treated as best estimates. Note that the vertical axis is in millions of dollars.

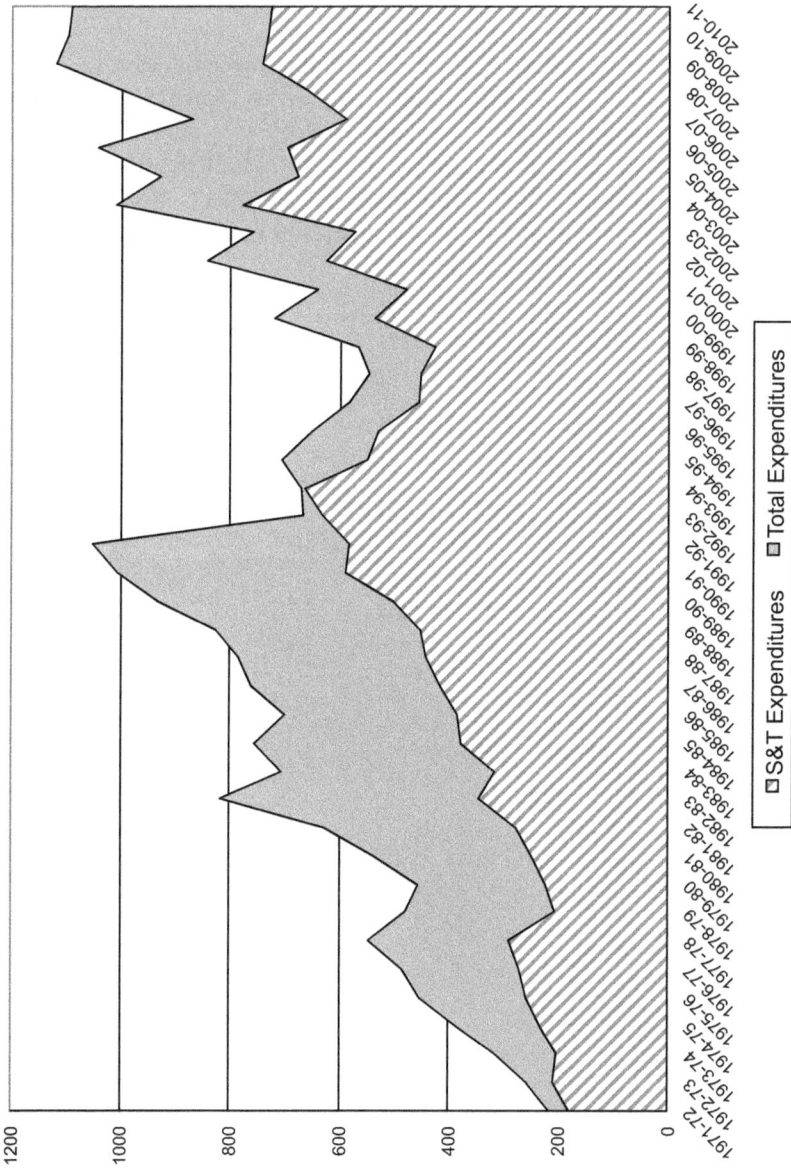

Appendix 2: Personnel[1]

EC Person Years – Total and S&T, 1971-72 to 2010-11

This chart shows that the majority of EC employees during its first 40 years were in S&T occupations: researchers, scientists, engineers, technicians or technologists. The data used here are derived from two different sources. The number of departmental employees is taken from reports to Parliament, of S&T employees from Statistics Canada. In both cases, the table uses the *actual* number of person years for any given year.[2]

---

[1] My thanks to Matthew Wallace for designing the charts in this appendix.
[2] A "person year" is the equivalent of the standard amount of work done by one person in a year.

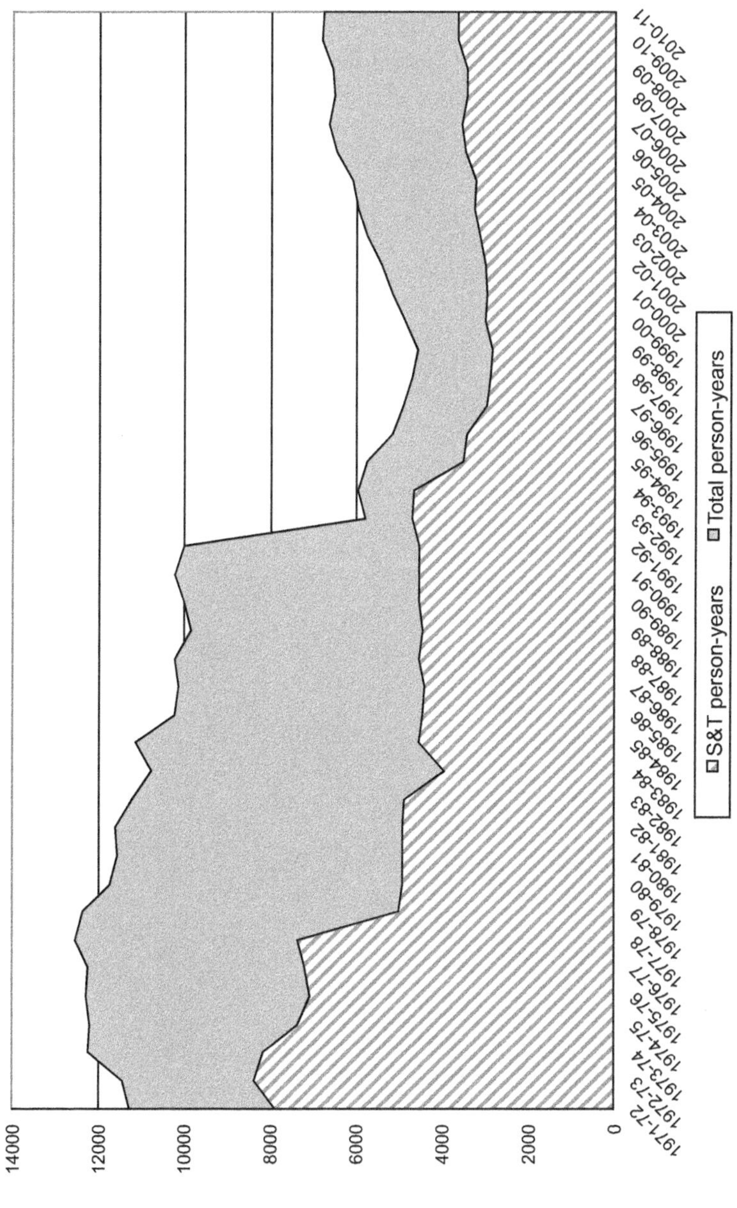
14000
12000
10000
8000
6000
4000
2000
0
1971-72
1972-73
1973-74
1974-75
1975-76
1976-77
1977-78
1978-79
1979-80
1980-81
1981-82
1982-83
1983-84
1984-85
1985-86
1986-87
1987-88
1988-89
1989-90
1990-91
1991-92
1992-93
1993-94
1994-95
1995-96
1996-97
1997-98
1998-99
1999-00
2000-01
2001-02
2002-03
2003-04
2004-05
2005-06
2006-07
2007-08
2008-09
2009-10
2010-11
S&T person-years
Total person-years

EC's Overall and S&T person-years

Percentage of Federal S&T Person Years in EC, 1971-72 to 2010-11

This chart shows the size of the EC S&T workforce in comparison to the total number of S&T staff employed by the federal government. The number of federal and of EC S&T person years are taken from Statistics Canada publications.

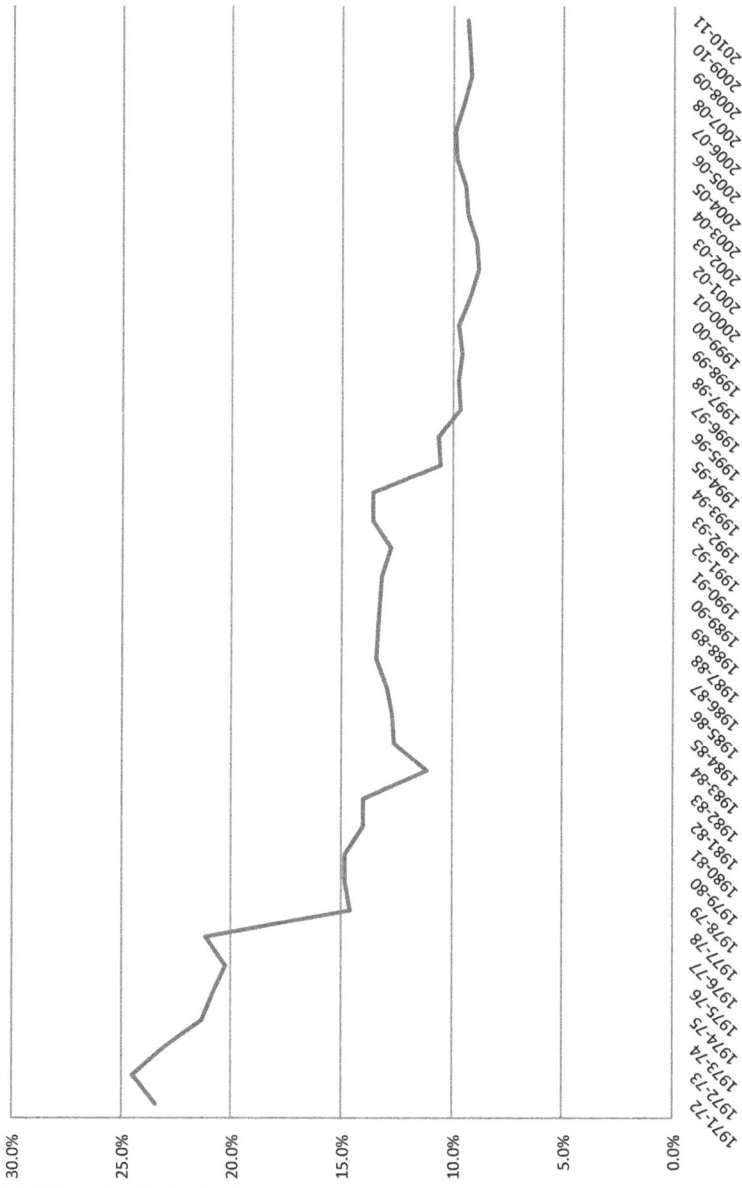

EC S&T person-years as a percentage of total federal S&T person-years

242

Appendix 3: Research Publications[1]

Number of EC Research Publications, 1971-2010

This chart shows the number of research publications by EC staff appearing in peer-reviewed journals during the Department's first 40 years. The Department made a significant contribution to advancing environmental knowledge both in Canada and internationally. The data are taken from Science Reuter's Web of Science.

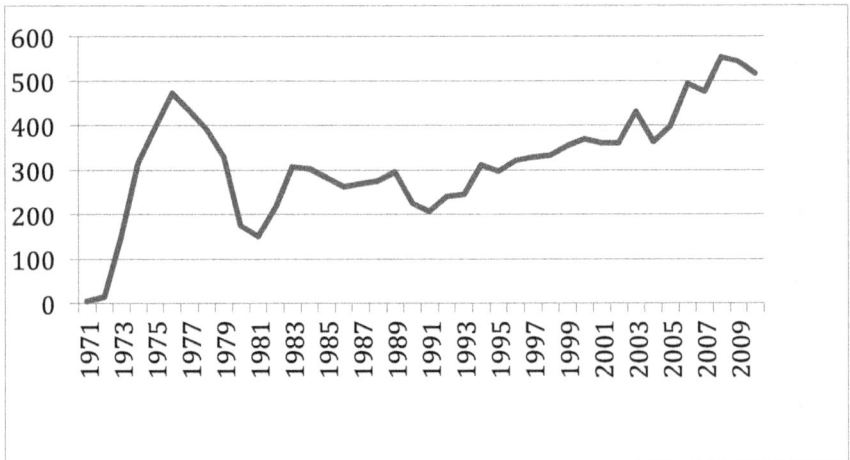

---

[1] My thanks to Yves Gingras for providing the information and graph.

Appendix 4: Timeline

The following table lists selected events mentioned in *Environment for Science* in chronological order. Its purpose is to help integrate the book's thematic organization by situating events relative to others occurring in the same time period.

1970	➤ Speech from the Throne (Pierre Trudeau's Liberal Government) announces intention to create a department of the environment – October ➤ Groups from other departments transferred to Fisheries & Forestry in preparation for the new department – November
1971	➤ Environment Canada officially established with Robert Shaw as DM – June ➤ Research Coordination is a directorate in the Policy, Planning & Research Service – June ➤ MOSST created – August
1972	➤ *Science in a Changing Environment: Proposals for a Departmental Science Policy* – February ➤ Federal Make-or-Buy Policy approved – February ➤ Report of the Cross-Mission Task Force on Research Facilities – March ➤ Working group on water analytical labs established – April ➤ Kenneth Hare appointed DG of Research Coordination – May ➤ Science Subvention Policy approved – August
1973	➤ EC reorganization: Research Coordination becomes Research Policy Directorate – January ➤ DGs Science Committee formed (lasts about 2 years) – February ➤ Office of the Science Advisor created with Hare as Science Advisor – August ➤ Fred Roots becomes Science Advisor – December
1974	➤ Office of the Science Advisor assigned responsibility for energy policy in EC – January ➤ Minister of State for Fisheries appointed – August ➤ Blair Seaborn starts as DM – December
1975	➤ 1st meeting of the Interdepartmental Committee on International S&T Relations – March ➤ *The Role of the Federal Government in S&T: A Conceptual Framework* – May ➤ 1st meeting of the Subcommittee on Research, Interdepartmental Committee on Environment – November ➤ *Report on the Development of Scientific & Professional Staff for Management* – November

1976	➤ EC presentation to Senate Special Committee on Science Policy – August ➤ EC renamed Department of Fisheries and Environment – September (until April 1979)
1977	➤ *An Overview of International Intergovernmental Environmental Relations* – April ➤ DM announces to departmental management committee the government's intention to create a Department of Fisheries & Oceans – September ➤ Minister of State for Environment appointed – September
1978	➤ Government decides to enhance technology transfer from government labs – April ➤ Government announces intention to privatize Forest Products Laboratories – September ➤ EC Policy Directive on Technology Transfer – December
1979	➤ Fisheries leaves EC; staff of Office of Science Advisor transferred to new Corporate Planning Group; Science Advisor reports to DM – April ➤ Parks Canada transferred into EC – June ➤ Joe Clark becomes Prime Minister, leading a Progressive Conservative government – June ➤ *A Scientific Scream* – October ➤ OAG criticizes EC management of science – December
1980	➤ DM announces Science Review – February ➤ Pierre Trudeau becomes Prime Minister, leading a Liberal government – March ➤ Minister John Roberts responsible for both EC & MOSST (to August 1983) – March ➤ CEAC report on EC's research policies and priorities – June ➤ Science Plan Action Plan approved – September
1981	➤ Report of ADM ECS on analytical labs – September
1982	➤ EC RES promotion procedures revised – November ➤ Jacques Gérin starts as DM – December
1983	➤ Report of Task Force on Environmental Protection Technologies – February ➤ *Maintenance & Utilization of Science in the Department of the Environment* – September
1984	➤ John Turner becomes Prime Minister, leading a Liberal government – June ➤ Report of Task Force on Federal Policies and Programs for Technology Development (Wright) – July ➤ Forestry Service moved from EC – September ➤ Brian Mulroney becomes Prime Minister, leading a Progressive Conservative government – September ➤ *Policy respecting S&T* – October

1985	➤ 1st meeting of Senior Scientists Committee (ceased meeting a year later) – May
	➤ Geneviève Ste-Marie starts as DM – August
1986	➤ Report of task force on management of S&T – January
	➤ Brundtland Commission visits Canada – May
	➤ Environmental Conservation Service and Environmental Protection Service merged – November
	➤ 1st meeting of the Working Group on Science Management – November
1987	➤ Report on Corporate Coordination of S&T at EC – January
	➤ 1st meeting of NABST – February
	➤ Federal *Decision Framework for S&T* – March
1988	➤ *Inventory of Canadian Environmental Industries* – February
	➤ EC National Workshops on S&T – September
1989	➤ Len Good starts as DM – May
	➤ Fred Roots retires as Science Advisor, becomes Science Advisor Emeritus – June
	➤ S&T Strategy workshop – July
	➤ Office of the Science Advisor recreated with Alex Chisholm as Science Advisor – November
1990	➤ Senior managers retreat launches EC transition exercise – July
	➤ Proposed Board on Environmental Research & Technology rejected – October
	➤ Green Plan released – December
1991	➤ Arctic Environmental Strategy – April
	➤ *R&D Profile and Issues Concerning the Delivery and Management of Science* – June
	➤ GOCO experiment on Wastewater Technology Centre commences – July
	➤ Eco-Research program announced – September
	➤ Technology for Solutions announced – October
1992	➤ *A Compendium of R&D in Environment Canada* – October
	➤ 1st EC Science Forum – November
1993	➤ Kim Campbell becomes Prime Minister, leading a Progressive Conservative government – June
	➤ Nick Mulder starts as DM – June
	➤ Parks Canada moved from EC – June
	➤ Environmental Conservation and Protection Service split into two Services – October
	➤ Office of the Science Advisor becomes Science Policy Branch in the Environmental Conservation Service – October
	➤ Jean Chrétien becomes Prime Minister, leading Liberal government – November
1994	➤ Government announces Program Review and Federal S&T Review – February

	➤ Mel Cappe starts as DM – May
	➤ 2nd EC Science Forum – May
	➤ Canadian Environmental Industries Strategy announced – September
	➤ Action Plan for managing S&T at EC – November
1995	➤ S&T Executive Committee formed – January
	➤ MOU on S&T for Sustainable Development established – January
	➤ EC announces Program Review decisions – February
	➤ Consultant's suggestions for an EC S&T communications strategy – March
	➤ EC puts in place business lines – April
	➤ 1st meeting of the interdepartmental ADM Steering Committee for S&T Human Resources – November
1996	➤ Federal S&T Strategy (*S&T for the New Century*) published – March
	➤ Interdepartmental ADMs Committee on Northern S&T – May
	➤ Ian Glen starts as DM – August
1997	➤ 1st episode of Earth Tones broadcast on Discovery Channel – January
	➤ Clerk launches La Relève – February
	➤ 1st meeting of R&D Advisory Board – April
	➤ 1st meeting of interdepartmental ADMs Committee on science in government – August
1998	➤ 1st meeting of federal Council of S&T Advisors – June
	➤ Len Good starts second term as DM – September
	➤ Toxics Substances Research Initiative announced – fall
	➤ Interdepartmental science communications action plan approved – November
	➤ EC workshop on training scientists in communications – November
	➤ 1st Federal Science Managers Forum – December
1999	➤ Collaborative S&T Positions Policy – January
	➤ Science Advice for Government Effectiveness released – June
	➤ Alan Nymark starts as DM – October
2000	➤ *S&T Partnering: Principles and Practices* – February
	➤ Budget announces Sustainable Development Technology Foundation – February
	➤ Science communications framework for EC – March
	➤ Federal Framework for S&T Advice released – May
	➤ Business line research agendas prepared – summer
	➤ Federal Northern S&T framework and action plan released – August
2001	➤ Workshop on a Canadian Environmental Sciences Network – January
	➤ Report on EC workshops on S&T values & ethics – March

	➤ Atlantic Environmental Sciences Network launched – May
	➤ Work starts on proposal for FINE (Federal Networks of Centres of Excellence) – September
2002	➤ 2nd Federal S&T Forum – October
	➤ National Wildlife Research Centre building completed at Carleton University – October
	➤ Science Policy Branch reports to ADM-ECS – December
2003	➤ Suzanne Hurtubise starts as DM – June
	➤ Establishment of S&T ADMs Integration Board – June
	➤ Paul Martin becomes Prime Minister, leading a Liberal government – December
2004	➤ TBS creates Science Infrastructure Review Working Group – spring
	➤ Arthur Carty named National Science Advisor – April
	➤ Samy Watson starts as DM – May
	➤ S&T Management Review Panel report – November
2005	➤ 3rd Federal S&T Forum – January
	➤ Beyond the Horizon workshop – September
	➤ Federal funding announced for International Polar Year 2007-08 – September
	➤ S&T Branch created; Science Policy transferred to the Branch – fall
2006	➤ Stephen Harper becomes Prime Minister, leading a Conservative government – February
	➤ New RES Career Progression Framework – February
	➤ Michael Horgan starts as DM – May
	➤ TBS creates Federal Laboratory Infrastructure Project – summer
2007	➤ Budget commitment to create an independent expert panel on the transfer of federal non-regulatory labs – March
	➤ Science Plan published – spring
	➤ Science ADMs Advisory Committee on Human Resources merged with Integration Board – April
	➤ Federal S&T Strategy (*Mobilizing S&T to Canada's Advantage*) released – May
	➤ Government announces intention to build an Arctic Research Station – October
2008	➤ New EC media relations policy – February
	➤ *Inter-Sectoral Partnerships for Non-Regulatory Federal Laboratories* – June
	➤ Ian Shugart starts as DM – August
2009	➤ Budget provides funding for federal labs and Arctic research facilities – January
	➤ EC's technology role approved by departmental management committee – February
2010	➤ *Environment Canada's International Polar Year Achievements* – June
	➤ Paul Boothe starts as DM – July

Appendix 5: Abbreviations

5NR	Five Natural Resource Departments
AAFC	Agriculture and Agri-Food Canada
Acc.	Accession
ADM	Assistant Deputy Minister
APEX	Association of Professional Executives of the Public Service of Canada
C&P	Conservation and Protection Service
CCMD	Canadian Centre for Management Development
CFI	Canada Foundation for Innovation
CIHR	Canadian Institutes for Health Research
CPG	Corporate Planning Group
CRTI	Chemical, Biological, Radiological-Nuclear, and Explosives Research and Technology Initiative
DFO	Department of Fisheries and Oceans
DG	Director General
DM	Deputy Minister
DOE	Department of the Environment
EC	Environment Canada
EMS	Environmental Management Service
FINE	Federal Innovation Networks of Excellence
IRAP	Industrial Research Assistance Program
LAC	Library and Archives Canada
MC	Memorandum to Cabinet
MOSST	Ministry of State for Science and Technology
MOU	Memorandum of Understanding
NATO	North Atlantic Treaty Organization
NHRI	National Hydrology Research Institute
NGO	Non-Governmental Organization
NRC	National Research Council
NRCan	Natural Resources Canada
NSERC	Natural Sciences and Engineering Research Council of Canada
NWRI	National Water Research Institute
O&M	Operations and Maintenance
OSA	Office of the Science Advisor
R&D	Research and Development
RES	(Classification used for Research Scientists)
RG	Record Group
RSA	Related Scientific Activities (non research scientific activities)
S&T	Science and Technology
SBDA	Science-based Department and Agency
SSHRC	Social Sciences and Humanities Research Council of Canada
TB	Treasury Board
TBS	Treasury Board Secretariat

Appendix 6: Note on Sources & Organizational Names

This book is based on my reading of a great many primary documents. Unfortunately, there are few secondary sources. Environment Canada has not been the subject of much historical scholarship – a history of the Department still awaits an author.[1] In addition, the Department's efforts in policy for science are rarely noted in works about the history of Canadian S&T (itself a relatively neglected area). For that reason, the book contains no bibliography. Instead, readers interested in following up on various events or themes should commence with references contained in the footnotes.

Almost all of the documents I consulted are held in three locations. One is Library and Archives Canada. The acronym *LAC* in footnotes indicates that that institution holds the source material. If no record group is listed, then the document comes from Record Group 108, the collection of EC papers in Library and Archives. Accession numbers and box numbers are listed. Finding aids in Library and Archives Canada should help in locating specific documents in their boxes.

The second is EC's records management services. Most of the documents I used were originally held in the Science Policy Division, and cover the period from approximately the mid-1980s to the present. They were collected by me during my years working in science policy in the Department and were transferred to records management in September 2010, just before I retired. This collection is organized into a number of groups of boxes according to their origin – e.g., the office files of Fred Roots, or the S&T Management Committee. Documents from this collection are indicated by *EC* in the footnotes, followed by the name of the group – e.g., EC Roots, or EC S&T Management Committee – and box number. File lists exist for all of these boxes. In the normal course of events, files held by EC's records management are transferred to Library and Archives Canada, after five years or more, for its decision on what to preserve. A small number of paper documents remain in the Science Policy Division due to relevance to its current work. And all electronic files are kept by the Division. This material is referred to in the footnotes as *Science Policy Division files*.

---

[1] The closest to a departmental history is G. Bruce Doern & Thomas Conway, *The Greening of Canada: Federal Institutions and Decisions* (Toronto: University of Toronto Press, 1994). Its focus is on environmental policy.

The final location is the Environment Canada library. It holds many of the grey literature reports prepared by or for the Department – e.g., the Science Policy Branch working paper series. Documents in the footnotes that are not attributed to LAC, EC or the Science Policy Division, can usually be found in the departmental library and sometimes in other libraries.

Although documents are the main source for this book, I have consulted with some former employees about their memories of events covered here. These consultations were not extensive. Information from them is indicated in the footnotes as a *personal communication*.

The names of organizational units in EC have often changed over the 40 years covered by this history. Units are referred to, in passages of this book, by the name in use at the time being dealt with. There is one exception to this. The Department's management committee has been known by such names as the Senior Management Committee and the Environment Management Board, among several others. To minimize confusion, I have used "departmental management committee" throughout the book. Readers should also note that for most of its history, the hierarchy of the Department was made up of Services (headed by ADMs), Directorates (by DGs), Branches (by directors) and Divisions (by chiefs). As a result of the 2005 reorganization, Services were renamed Branches and the former branches and divisions became Divisions and Sections, respectively.

Index

www.ingramcontent.com/pod-product-compliance
Lightning Source LLC
Chambersburg PA
CBHW030005290326
41934CB00005B/229